Manufactured Carbon: A Self-Lubricating Material for Mechanical Devices

Author

R. Robert Paxton

Chief Engineer
Pure Carbon Company
St. Marys, Pennsylvania

CRC Press
Taylor & Francis Group
Boca Raton London New York

CRC Press is an imprint of the
Taylor & Francis Group, an informa business

First published 1979 by CRC Press
Taylor & Francis Group
6000 Broken Sound Parkway NW, Suite 300
Boca Raton, FL 33487-2742

Reissued 2018 by CRC Press

Library of Congress Cataloging in Publication Data

Paxton, Ralph Robert, 1920-
Manufactured carbon.

Bibliography: p.
Includes index.
1. Carbon. I. Title.
TA455.C3P39 622'.92 78-10107
ISBN 0-8493-5655-5

A Library of Congress record exists under LC control number: 78010107

Publisher's Note
The publisher has gone to great lengths to ensure the quality of this reprint but points out that some imperfections in the original copies may be apparent.

Disclaimer
The publisher has made every effort to trace copyright holders and welcomes correspondence from those they have been unable to contact.

ISBN 13: 978-1-315-89518-5 (hbk)
ISBN 13: 978-1-351-07428-5 (ebk)

Visit the Taylor & Francis Web site at http://www.taylorandfrancis.com and the
CRC Press Web site at http://www.crcpress.com

PREFACE

This book took more than five years of research and writing. It could not have been done without encouragement and help from my family and business associates. I particularly want to thank Mary Lou and Anne for tiptoeing about the house on weekends while I worked, Nancy L. Paxton for assistance with the copy editing, Edward Strugala for the graphs and line drawings, Howard Hulburt for many of the photographs and for numerous technical suggestions, and Betty Aaron and Marlene Lenze for typing and retyping the evolving manuscript.

I also want to thank my employer, Pure Carbon Company, Inc., for their support and for permission to use hitherto unpublished data.

R. R. Paxton

THE AUTHOR

R. Robert Paxton, Sc.D. has been Chief Engineer for the Pure Carbon Company, St. Marys, Pennsylvania since 1958. During much of this time he was responsible for new product development, product testing, process development, customer technical service, manufacturing technical service, and corporate planning.

Dr. Paxton received his B.S. degree in the field of chemical engineering from the University of Illinois in 1943. In 1949 he received his Sc.D. in chemical engineering from the Massachusetts Institute of Technology for his research in carbon technology. Before joining Pure Carbon Company, Dr. Paxton was Assistant Professor of Chemical Engineering at Stanford University.

Dr. Paxton is a licensed engineer in the states of California, Illinois, and Pennsylvania. He is a member of the American Institute of Chemical Engineers, the American Chemical Society, the American Society of Lubrication Engineers, and the Electrochemical Society. From 1968 to 1977 he was Secretary and Vice Chairman of the Manufactured Carbon Division (C-5), of ASTM.

In the performance of his technical activities, Dr. Paxton has authored or co-authored a score of technical papers on the self-lubricating properties of carbon. He is also the inventor or co-inventor of five patents in this field.

TABLE OF CONTENTS

Chapter 1

INTRODUCTION

I. SCOPE

The purpose of this book is to provide engineering guidance to those who design and use manufactured carbon parts as rubbing members in mechanical devices. As an engineering guide, it will discuss economic and technical factors important in the proper use of mechanical carbon components. The applications discussed will range from mechanical seals to sleeve bearings and from piston rings to valve seats. Electrical applications (such as brushes for motors) and structural applications (such as rocket nozzles) are outside the scope of this book.

II. PURPOSE

There is no text presently available which describes the design and application of manufactured carbon to mechanical devices. Moreover, carbon is rarely mentioned in engineering textbooks or design manuals. Designers and users of mechanical carbon are, therefore, handicapped in the proper application of this material. The net result is an economic loss due to improper design or, more likely, use of a less desirable alternate material.

III. SOURCE OF INFORMATION

In the past 30 years a large amount of information has been published on both the uses and the performance of carbon in mechanical devices. The four major sources of information are the technical literature, government reports, patents, and private company publications. All these sources will be drawn upon as needed to help the designer and user of carbon parts. Supplementing these will be the engineering experience of the author and his co-workers.

IV. NEW DEVELOPMENTS

The use of carbon in mechanical devices is expanding at the rate of 5 to 15% a year. This rate of growth is created by extensive research, development, and testing in both producer and user laboratories. As a result, new products and new applications quickly augment or replace the old. Thus, the designer should consult current supplier catalogs for information on the newest materials available. If the designer has any questions about the application of carbon to his design, he should contact a supplier to obtain information on the latest design advances. The major suppliers provide a sizeable engineering group for the sole purpose of assisting users.

V. HISTORICAL PERSPECTIVE

Carbon has long been used by industry as a fuel and as a reducing agent in metallurgical operations. Its use as an engineering material is, however, quite new. In 1906, the first account of its use was published by Salto[1] who used it as a current collector (brush) in an electric motor. He noted the unique self-lubricating properties of carbon along with its ability to transfer an electrical current to a moving conductor. By 1970 the manufacture of carbon brushes was a $30 million a year business in the U.S., and several books[2-4] were devoted to the design and application of brushes.

Carbon was probably first used as a purely mechanical component in seal rings in steam turbines around 1910. Scott and Deats[5] obtained a patent on metal-impregnated carbon bearings in 1913. The next application to involve a significant volume was for piston rings in nonlubricated gas compressors. The overall use of carbon remained small, however, and the growth rate of mechanical carbon applications was low until the early 1950s.

The introduction of jet aircraft and the rapid expansion of commercial aviation changed this. Concurrently the chemical and petroleum industries also entered a period of rapid expansion. As a consequence, there was a large increase in the demand for mechanical seals. Carbon quickly gained acceptance as a rubbing element in mechanical seals,[6] and the mechanical applications of carbon entered a period of rapid growth. If present trends continue, the sale of mechanical carbon components in the U.S. will surpass that of carbon brushes before 1985.

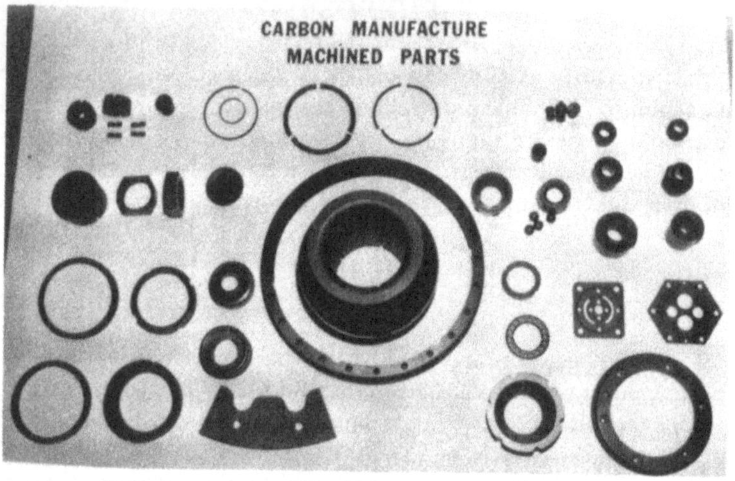

FIGURE 1.1. A typical collection of manufactured carbon parts used as mechanical components.

REFERENCES

1. **Salto, S.**, *Elektrotech, Z.*, September 20, 1906.
2. **Shobert, E. I., II**, *Carbon Brushes*, Chemical Publishing, New York, 1965.
3. **Neukirchen, J.**, *Carbon Brushes* (trans. by Shobert, E. I., II), Telegraph Press, Harrisburg, Pa., 1937.
4. **Holm, R.**, *Electic Contacts Handbook*, Springer-Verlag, Berlin, 1958.
5. **Scott, C. and Deats, W.**, U.S. Patent 1,053,881, 1913.
6. **Beacham, T. E.**, Rotary and oscillating seals, *Engineering (London)*, 187, 228, 1949.

Chapter 2

PROPERTIES OF GRAPHITE CRYSTALS

I. INTRODUCTION

Carbon, the sixth element in the atomic series, is the 19th most abundant on the surface of the earth. It has four valence electrons, all of which can enter into σ-type or covalent bonds. The valence electrons of carbon can also form less energetic π-type or resonant bonds. This arrangement of valence electrons gives carbon the ability to form compounds with many other elements, compounds which vary widely in properties. Indeed, the variety of carbon compounds is so vast that an entire field of science, organic chemistry, has arisen from their study.

II. CRYSTALLINE CARBON

Our interest, however, is in "pure" carbon, carbon materials where the noncarbon elements play only a supporting role. An understanding of manufactured carbon can best be developed by first reviewing the properties of the element.

Elemental carbon exists in two crystalline forms at room temperature, diamond and graphite. The denser diamond form is metastable at normal pressures and temperatures. Its rate of reversion to graphite becomes measurable when the temperature exceeds 1370°C (2500°F).

From an engineering standpoint, the two most important consequences of the atomic structure of carbon are its low density and its ability to form energetic bonds. The hardness of the diamond and the high melting point of graphite are two direct consequences of the latter quality.

Table 2.1 lists some properties of graphite crystals. I have chosen the values that are shown from recent literature, but it must be pointed out that reported values vary widely. Palmer and Shelef,[1] for example, reviewed a score of carbon melting point "determinations" made over the last 40 years. They found carbon melting points ranging from 3200 (5792°F) to 7425°C (13,397°F), and then concluded the probable value was between 4325 (7817°F) and 4525°C (8177°F). The value of 4375°C (7907°F) recorded in Table 2.1 is from Fateeva.[2]

Vapor pressure, like melting point, is difficult to measure owing to the high temperatures involved. The values reported are from the "approximate" curves of Palmer and Shelef.[1] The same authors calculated the heat of vaporization (sublimation) from spectrographic data on the disassociation of carbon monoxide together with appropriate thermochemical data from the JANAF[3] compilation.

The specific heat of graphite increases linearly with increasing temperature from a value of 1.3 kJ/kg atom °C at −193°C (0.025 Btu/lb °F at −315°F) to 13.8 kJ/kg atom °C at 177°C (0.275 Btu/lb °F at 350°F). It continues to rise at higher temperatures approaching the Dulong and Petit value of 26 kJ/kg atom °C (0.525 Btu/lb °F) asymptotically.[4]

The graphite crystal may be visualized as infinite parallel layers of hexagons stacked 3.4Å (3.4×10^{-10}m) apart. Figure 2.1 gives the principal crystal dimensions as reported by Franklin[5] and Bacon.[6] Because of this structure, scalar properties of graphite crystals vary markedly, depending upon whether they are measured parallel or normal to the layer planes. Electrical resistance, for example, varies at least 100-fold, and the variation in thermal conductivity is only slightly less.

Thermal expansion is particularly interesting in that the room temperature value is negative parallel to the planes but strongly positive normal to the planes. It has been

TABLE 2.1

Properties of Graphite Crystals

Property	Metric (SI) units	U.S. customary units
Atomic weight	12.01	12.01
Density at 20°C (68°F)	2,268 kg/m^3	141.5 lb/ft^3
Melting point at 100 atmospheres	4,375°C	7907°F
Vapor pressure at		
2,204°C (4,000°F)	0.1 N/m^2	10^{-6} at
2,760°C (5,000°F)	100 N/m^2	10^{-3} at
3,316°C (6,000°F)	7×10^4 N/m^2	0.7 at
3,871°C (7,000°F)	6×10^5 N/m^2	6 at
Heat of vaporization	710 MJ/kg atom	25,500 Btu/lb
Specific heat at 1 atmosphere at		
20°C (68°F)	8.5 kJ/kg atom °C	0.17 Btu/lb °F
260°C (500°F)	16.5 kJ/kg atom °C	0.33 Btu/lb °F
537°C (1,000°F)	20.6 kJ/kg atom °C	0.41 Btu/lb °F
1,097°C (2,000°F)	23.0 kJ/kg atom °C	0.46 Btu/lb °F
2,205°C (4,000°F)	26.0 kJ/kg atom °C	0.52 Btu/lb °F
Electrical resistance at 20°C (68°F)		
Parallel to layer planes	4×10^{-7} Ω m	1.6×10^{-5} Ω in.
Normal to layer planes	4×10^{-5} Ω m	1.6×10^{-3} Ω in.
Thermal conductivity at 20°C (68°F)		
Parallel to layer planes	400 W/m °C	232 Btu/ft °F hr
Normal to layer planes	6 W/m °C	3.5 Btu/ft °F hr
Coefficient of thermal expansion (average) 20—537°C (68—1,000°F)		
Parallel to layer planes	-0.36×10^{-6} cm/cm°C	-0.2×10^{-6} in./in. °F
Normal to layer planes	30×10^{-6} cm/cm °C	16.6×10^{-6} in./in. °F

hypothesized that the lengthening carbon-carbon bonds cause "wrinkling" of the planar hexagons to give a net shrinkage of this plane.

Perfect crystals of graphite are as rare as diamonds of similar size. Crystallites of 10^7Å (1 mm) in length, for example, would be considered unusually large.[7] Fortunately, the crystallites in naturally occurring graphite flakes are often well aligned; hence, it is possible to make scalar measurements on polycrystalline flakes and get values which approximate those for single crystals.

As a rule, the crystallites in manufactured carbon range from 10 to 10,000 Å. They are generally too small to be resolved by a metallurgical microscope. Individual filler grains, however, can be readily seen. These generally range from 250,000 to several million angstroms.

Graphite is a birefringent crystal and appears to change color under polarized light as the microscope stage is rotated. Alignment of the crystallites in a filler grain or flake is often good enough to cause the entire grain to look like a single crystal. This type of microscopic examination is often used to estimate the concentration of larger graphite crystallites in a manufactured carbon part.

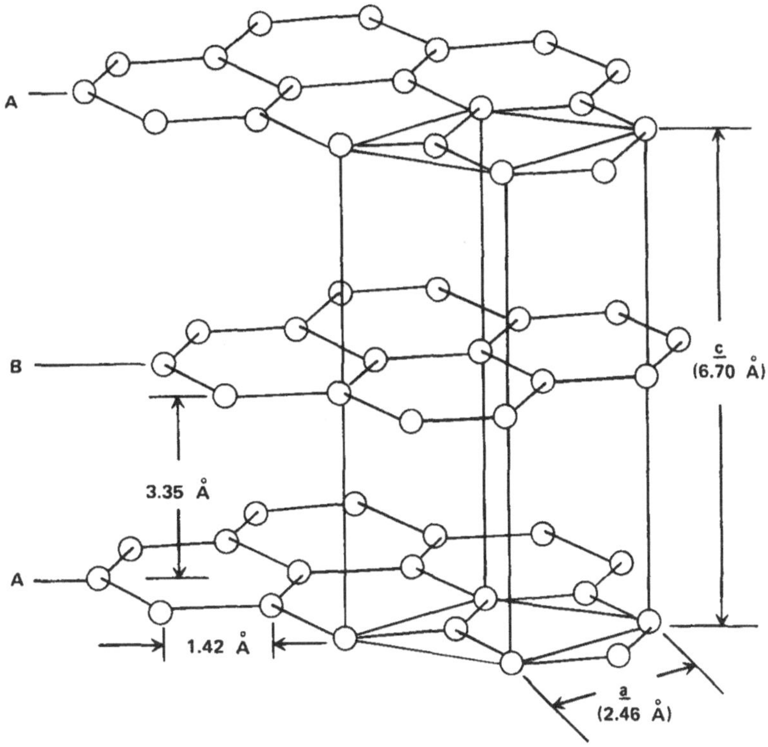

FIGURE 2.1. Graphite crystal structure.

III. CARBON AS A SELF-LUBRICATING SOLID

The key to the value of carbon as a self-lubricating solid lies in its laminar crystal structure, its ability to form strong chemical bonds with gases such as water vapor, and the fact that adsorbed gas weakens the interlayer bonding forces. An understanding of these factors is important when selecting carbon grades for applications in mechanical devices.

Carbon adsorbs and holds polar gases (H_2O, O_2, CO_2, NH_3, CH_3OH, etc.) tenaciously. Lang and Magnier,[8] for example, found that prolonged evacuation at 975°C (1787°F) fails to remove all adsorbed gases. Carbon which has been degassed at 2725°C (4937°F) quickly readsorbs oxygen and water vapor when cooled and exposed to air. Thus, all polycrystalline carbon bodies that one is likely to encounter contain large amounts of adsorbed gas in their structure.

Adsorption takes place at the edges of the crystallites where there are unsatisfied bonding forces. Small crystallites have a higher concentration of edge atoms — and hence adsorption sites — than large crystallites. Thus, carbon bodies produced from fillers having a high concentration of small crystallites tend to adsorb more gases.

Adsorbed polar gases aid in reducing rubbing friction by reducing the interlayer bonding forces of the graphite crystal. In his experiments, Bryant et al.[9] demonstrated that the work of cleaving graphite layers at room temperatures was reduced from 1750 to 200 ergs/cm^2 when he admitted air into his apparatus. If he had been able to completely degas his specimen before cleaving in a vacuum, the work would have more closely approximated the theoretical cohesive energy of 2500 ergs/cm^2. Since rubbing involves the sliding of one surface past another, it follows that the absence of adsorbed gas should increase friction and wear of carbon bodies.

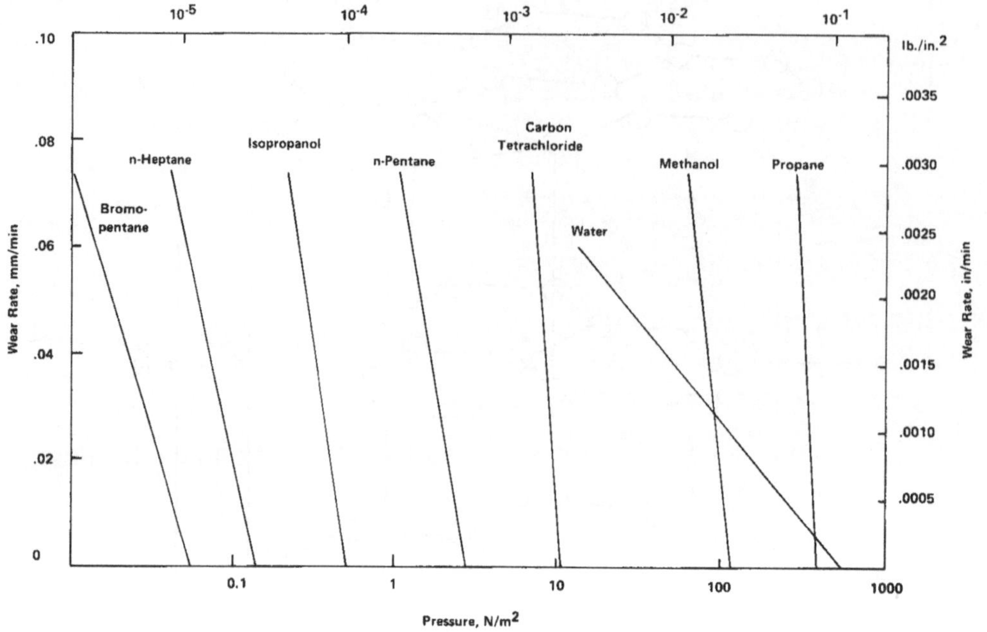

FIGURE 2.2. Effect of vapors on wear rate of graphite. (From Savage, R. H. and Schaefer, D. L., *J. Appl. Phys.*, 27, 136, 1956. With permission.)

The practical consequence of this characteristic of graphite was impressed upon users in the early 1940s. Generator brushes on newly developed high-flying bombers wore out in a few hours rather than lasting the normal hundreds of hours. Savage[10] duplicated this phenomena in the laboratory and reported that, in the absence of water vapor, graphite had no lubricating properties. Many others[11-19] have contributed to a better understanding of this phenomenon, and their findings are reviewed in six books[20-25] published since 1963.

Water vapor is not the only adsorbable gas which enhances the lubricating property of graphite. Figure 2.2 shows the reduction in wear rate brought about by other vapors, some of which are effective at extremely low partial pressures relative to their vapor pressure at room temperature.

Many inorganic additives such as lead iodide, silver iodide, silver sulfide, calcium carbonate, barium fluoride, cadmium bromide, silver metal, and others[26-30] have also been found to be very effective in reducing friction and wear where adsorbable gases are absent. Additives are widely used in manufactured carbon parts such as seals and bearings destined for use in applications where adsorbable gases may be in short supply.

IV. FILM FORMATION BY CARBON BODIES

In most engineering applications the carbon part rubs against a smooth metal surface. Close examination of this metal surface after the device has been operated for an extended period of time reveals a stained wear track. The build-up of solids in the wear track is generally 2000 to 10,000 Å (0.2 to 1.0 × 10^{-6} m) deep, which is thick enough to see but not thick enough to measure with ordinary laboratory equipment. Paxton and Shobert[31] first reported the existence of this film on seal mating faces and speculated on its role in controlling friction and wear rates.

Buckley and Johnson[18] suggested this wear track was coated with graphite platelets bonded to the metal substrate through oxide linkages. They postulated a carbon-oxygen bond and noted a 1000-fold increase in wear rate when the air pressure surrounding the interface was reduced to 1.3×10^{-5} N/m² (1.9×10^{-9} psi). Concurrent with this increase in wear was disappearance of carbon from the wear track.

Experimental findings by other investigators[32-35] are so similar that there can be no doubt that the self-lubricating ability of carbon depends on the presence of foreign (i.e., noncarbon) aids. These aids weaken the carbon-carbon bonds near the rubbing surface, promote the formation of transfer films, or both. Nature usually provides these aids without help from man; hence, carbon is generally credited with being a self-lubricating solid. However, in applying carbon to mechanical devices, the designer must be ever aware of the need for adsorbable vapors and for a mating surface which promotes the formation of a transfer film. Lacking either of these, carbons with appropriate organic or inorganic additives should be used.

REFERENCES

1. **Palmer, H. B. and Shelef, M.,** *Chemistry and Physics of Carbon,* Vol. 4, Marcel Dekker, New York, 1968, 85.
2. **Fateeva, N. S., Vereshchagin, L. F., and Kolotygin, V. S.,** *Sov. Phys. Dokl.,* (English trans.) 8, 893, and 904, 1964.
3. **JANAF,** Thermochemical Tables, Thermal Research Laboratory, Dow Chemical Co., Midland, Mich., 1960—1966; JANAF, Thermochemical Tables First Addendum, Clearinghouse for Federal Scientific and Technical Information, Washington, D.C., August, 1966.
4. **Reynolds, W. N.,** *Physical Properties of Graphite,* Elsevier, New York, 1968, 64.
5. **Franklin, R. E.,** *Acta Crystallogr.,* 4, 253, 1951.
6. **Bacon, G. E.,** *Acta Crystallogr.,* 3, 137, 1950.
7. **Walker, P. L.,** *Am. Sci.,* p. 259, June 1962.
8. **Lang, F. M. and Magnier, P.,***Carbon,* 2, 7, 1964.
9. **Bryant, P. J., Gutshall, P. L., and Taylor, L. H.,** A study of mechanisms of graphite friction and wear, *Wear,* 7, 118, 1964.
10. **Savage, R. H.,** Graphite lubrication, *J. Appl. Phys.,* 19, 1, 1948.
11. **Campbell, W. E. and Kozak, R.,** Studies in boundary lubrication. III. The wear of carbon brushes in dry atmosphere, *Trans. ASME,* 70, 491, 1948.
12. **Savage, R. H. and Schaefer, D. L.,** Vapor lubrication of graphite sliding contacts, *J. Appl. Phys.,* 27, 136, 1956.
13. **Deacon, R. F. and Goodman, J. F.,** Lubrication by lamellar solids, *Proc. R. Soc. London Ser. A,* 243, 464, 1958.
14. **Rowe, G. W.,** Some observations on the frictional behavior of boron nitride and graphite, *Wear,* 3, 274, 1960.
15. **Mordike, B. L.,** The lubrication of solids at high temperatures, *Trans. ASLE,* 3, 110, 1960.
16. **Cannon, P.,** Mechanism of the vapor lubrication of graphite, *J. Appl. Phys.,* 34, 20, 1963.
17. **Reichenbach, G. S. and Pourney, J. L.,** The importance of humidity in friction measurements, *Lubr. Eng.,* 20, 409, 1964.
18. **Buckley, D. HV. AND Johnson, R. L.,** Mechanism of lubrication for solid carbon materials in vacuum to 10^{-9} millimeter of mercury, *Trans. ASLE,* 7, 97, 1964.
19. **Braithwaite, E. R.,** Friction and wear of graphite and molybdenum sulfide, *Sci. Lubr.,* 18(5), 17, 1966.
20. **Bowden, F. P. and Tabor, D.,** *The Friction and Lubrication of Solids, Part II,* Claredon Press, Oxford, 1963.
21. **Braithwaite, E. R.,** *Solid Lubricants and Surfaces,* Pergamon Press, New York, 1964.
22. **Bryant, P. J., Lavik, M., and Salomon, G.,** *Mechanism of Solid Friction,* Elsevier, New York, 1964.
23. **Blackman, L. C. F.,** *Modern Aspects of Graphite Technology,* Academic Press, London, 1970.

24. **Buckley, D. H.**, Friction, Wear and Lubrication in Vacuum, NASA SP-277, National Aeronautics and Space Administration, Scientific and Technical Information Office, Washington, D.C., 1971.
25. **Clauss, F. J.**, *Solid Lubricants and Self Lubricating Solids,* Academic Press, New York, 1972.
26. **Churchill, R. W., Jr.**, U.S. Patent 3,165,480, 1965.
27. **McCrary, M. R. and Ramadanoff, D.**, Canadian Patent 587,500, 1959.
28. **Oliver, C. S. and Haltner, A. J.**, U.S. Patent 3,173,045, 1965.
29. **Churchill, R. W., Jr.**, Canadian Patent 631,817, 1961.
30. **Ramadanoff, D.**, Canadian Patents 580,790 and 580,791, 1959.
31. **Paxton, R. R. and Shobert, W. R.**, Testing high speed seal carbons, *Trans. ASLE,* 5, 308, 1962.
32. **Hady, W. F., Allen, G. P., and Johnson, R. L.**, Wear and friction of mechanical carbons in liquid oxygen as influenced by transfer films, *Trans. ASLE,* 6, 201, 1963.
33. **Bisson, E. E.**, Friction of metal, lubricating coatings and carbons in liquid nitrogen and hydrogen, in Advanced Bearing Technology, NASA SP-38, National Aeronautics and Space Administration, Scientific and Technical Information Office, Washington, D.C., 1964.
34. **Quinn, T. J. F.**, *Br. J. Appl. Phys.*, 15, 513, 1964.
35. **Grigorenko, L. PV., Plutova, L. A., and Rovinskii, B. M.**, *Sov. Phys. Dokl.*, 10, 154, 1965.

Chapter 3

PROPERTIES OF MANUFACTURED CARBON

I. INTRODUCTION

Manufactured carbon is a polycrystalline material whose structure contains many more defects and a wider range of crystallite sizes than naturally occurring graphite.

The range in crystallite sizes is largely controlled by the choice of raw materials. Wood charcoal, for example, has a crystallite size in the range of 50 Å (5×10^{-9} m). Parts made from manufactured carbon with a high content of wood charcoal tend to be hard and abrasive. They could be very reactive toward oxygen and other chemicals. Indeed, charcoal-based parts can have an internal micropore surface area exceeding 500 m^2/g and be so active chemically that they will spontaneously ignite if heated in air to 200°C.

At the other extreme are parts of manufactured carbon composed largely of synthetic or natural graphite. Average crystallite sizes here run to 1000 Å (100×10^{-9} m) and more. These parts are soft and unctuous with excellent lubricating properties but poor wear resistance. They are chemically inert, have micropore surface areas below 0.1 m^2/g, and can be heated in air above 700°C without igniting.

The two other major variables controlling the properties of manufactured carbon are the binder used and the firing temperature employed. These will be discussed in Chapter 4. First, however, let us look at the different types of manufactured carbon commercially available at present. Table 3.1 shows the properties of six molded carbons used as rubbing components in mechanical applications. These six were chosen to illustrate the more common types of carbon used in mechanical devices.

II. RESIN-BONDED CARBON

The first material, G-1, is a resin-bonded carbon typical of those used for light-duty seals and valve seats. The resin is a phenol formaldehyde, and the filler is finely ground flake graphite. Specific applications for this type of "carbon" include automobile water pump seals, seals for washing machines and other household appliances, and valve seats on gasoline metering pumps. Carbon G-1 cannot tolerate high rubbing temperatures or face loads above 0.35 MN/m^2 (50 $lb/in.^2$). Accordingly, its use is limited to applications where there is little chance of the rubbing area running dry for a prolonged period. Its principal advantage as an engineering material is that relatively complex, close-tolerance parts can be cheaply formed by compression molding in multicavity dies.

III. METAL-BONDED CARBON

The second manufactured carbon, S-95, is a silver-bonded graphite typical of those used for dry running bearings. This class of materials is noteworthy because it is unusually strong and yields and deforms significantly before failing in tension or flexure. It has been used with some success in slow-moving bearings which carry heavy loads and where misalignments in the system require a deformable bearing. Copper, like silver, does not react chemically with graphite at sintering temperatures; hence, it can be used as a metal binder. Copper graphites, however, are generally limited to temperatures below 175°C (347°F) because the copper phase oxidizes and loses its strength.

Metal-bonded carbons are generally considerably more expensive than resin-bonded carbons. The metal powders used in their manufacture are costly and much higher forming pressures are required. After pressing, they must be carefully sintered in neu-

TABLE 3.1

Properties of Molded Nonimpregnated Mechanical Carbons

Identification*	G-1	S-95	P-2W	P-9	P-5	P-03
Density (g/cc)	1.90	8.05	1.60	1.75	1.72	1.82
Open porosity (vol %)	10	3	25	16	18	8
Median pore diameter ($\mu m = 10^{-6}$ m)	0.12	40	3	0.7	1.5	1.0
Permeability (μDarcies)	10	10,000	25,000	1,000	6,000	1,000
Temperature limit in air						
°F	400	700	550	550	600	1,000
°C	204	371	288	288	316	538
Hardness (Scleroscope C)	20	10	45	65	82	75
Strength, MN/m^2 (psi)						
Compressive	41 (6,000)	210 (30,000)	83 (12,000)	145 (21,000)	207 (30,000)	140 (20,000)
Flexural	31 (4,500)	97 (14,000)	34 (5,000)	52 (7,500)	62 (9,000)	55 (8,000)
Tensile	21 (3,100)	55 (8,000)	28 (4,000)	28 (4,000)	41 (6,000)	38 (5,500)
Modulus of elasticity (psi)	3.4×10^6	2.0×10^6	1.4×10^6	2.1×10^6	2.4×10^6	1.8×10^6
(MN/m^2)	23×10^3	14×10^3	9.6×10^3	14×10^3	17×10^3	12×10^3
Coefficient of thermal expansion 68—600°F (20—314°C)						
(in./in. °F)	2.0×10^{-6}	1.2×10^{-6}	1.8×10^{-6}	2.4×10^{-6}	2.5×10^{-6}	2.0×10^{-6}
(mm/mm °C)	3.6×10^{-6}	2.0×10^{-6}	3.2×10^{-6}	4.3×10^{-6}	4.5×10^{-6}	3.6×10^{-6}
Thermal conductivity						
(Btu/hr ft °F)	1.0	80	10	7	5	40
(W/m °C)	1.7	140	17	12	9	70
Electrical resistivity						
(Ω-in.)	2.5×10^{-3}	0.5×10^{-6}	0.7×10^{-3}	$15. \times 10^{-3}$	1.8×10^{-3}	0.6×10^{-3}
(Ω-cm)	6.2×10^{-3}	1.5×10^{-6}	1.8×10^{-3}	3.8×10^{-3}	4.5×10^{-3}	1.5×10^{-3}
Anistropy	Negligible	Negligible	Negligible	Negligible	Negligible	Negligible

* Material designations are those of Pure Carbon Company, Inc., St. Marys, Pa.

tral or reducing atmospheres to develop proper knitting of the metal matrix and full strength in the part.

IV. LOW MODULUS CARBON

The third carbon material, P-2W, was developed for piston and wear rings on large, nonlubricated air compressors. It is made from relatively coarse synthetic graphite lubricating particles bonded with carbonized pitch. The high permeability of this type of carbon permits forming and baking large shapes without cracks or laminations caused by escaping gas when the binder is pyrolyzed.

Carbon P-2W is soft and easy to machine, yet it is unusually strong considering its porosity. In applications, its principal virtues are its low modulus and its high thermal conductivity. The former allows ready conformity to the compressor cylinder bore, and the latter keeps the rubbing interface cool.

Carbons such as P-2W require good finish on the mating metal surface. A surface finish coarser than 0.5 μm arithmetic average (AA) will abrade the carbon and lead to high wear rates. On the other hand, the mating metal surface need not be harder than a Rockwell C value of 30, providing the surrounding fluid is free of suspended hard particles. The synthetic graphite fillers used in carbons like P-2W generally have less than 0.5% ash; hence, the resultant carbons have little if any "polishing" action.

V. LOW PERMEABILITY, MACHINABLE CARBON

The fourth carbon grade shown, a low permeability, machinable carbon-graphite, was developed for sleeve bearings in submerged pumps. It is similar to the previous carbon except that a finely ground natural graphite is used, and the pitch to graphite ratio is higher.

The result is a harder, stronger material with a much finer pore structure. Despite this fine pore structure, billets as large as 14 inches square × 72 inches long (0.35 × 1.8 m) can be produced with baking cycles of 4 to 5 weeks. Smaller shapes can be produced with shorter cycles. The relatively low permeability of this carbon, P-9, is an aid to forming a hydrodynamic running film which reduces friction and wear in the wet bearing.

The natural graphite fillers used in carbons like P-9 generally contain 10 to 15% ash. Individual ash particles are usually smaller than 1 μm and are chalk-like or clayey. Nevertheless, they do impart "polishing" action to the resultant manufactured carbon. As a consequence, the carbon will slowly lap the rubbing metal surface; hence, initial metal surface finishes of 1 μm AA are perfectly suitable. Mating metal hardness should be 35 or higher on the Rockwell C scale.

In addition to their use in bearings, resin-impregnated versions of grades like P-9 are widely used as noses in mechanical seals. They have enough polishing action to keep the mating metal wear plate free of build-up from the sealed medium, yet they are gentle enough to be run against cast iron (Meehanite) or stainless steel (17-4 PH) without excessive wear.

VI. HIGH STRENGTH CARBON

The fifth carbon, P-5, was developed for applications where strength, stiffness, and abrasion resistance were the prime requirements. Finely divided calcined petroleum coke is used in the filler to achieve these properties. A typical application would be vanes for dry air pumps.

Particle size of the coke filler controls the abrasiveness of carbons like P-5. Calcined petroleum coke is essentially 100% carbon, yet it is hard enough to scratch most metals. By mixing powdered coke and powdered graphite in the proper proportions, one can formulate a carbon with many of the lubricating properties of P-9 combined with extremely good wear resistance.

Manufactured carbons which contain significant concentrations of hard carbons — calcined petroleum coke, charcoal, glassy carbon, etc. — require very hard mating surfaces. Typical choices would be chrome-plated steel, tungsten carbide, silicon carbide or fused alumina. These carbons have the ability to continuously polish the mating faces; hence, they can be used for seal noses where the fluid sealed tends to deposit tar or coke at the interface.

VII. HIGH THERMAL CONDUCTIVITY CARBON

Seal face rubbing speeds to 30,000 ft/min (150 m/sec) are encountered in aircraft jet engines. Frictional heat generated at the interface must be conducted away, and this requires a high thermal conductivity carbon. High strength, abrasion resistance, and oxidation resistance are also needed to give long operating life under the severe conditions encountered. The sixth type of carbon, P-03, was developed for these conditions. The finest available filler carbon, together with high pitch loadings, was used to obtain strength and abrasion resistance. Thermal conductivity was increased by graphitizing (heat treating) the carbon at temperatures above 2760°C (5000°F).

Generally speaking, heat treatment of carbons to high temperature does little to alter grain size or grain orientation of the crystallites in most filler materials. Crystallites in the carbonaceous residue of the binder can be slightly altered. The change, based on X-ray diffraction patterns, is at best a doubling of average crystallite size. Average carbon-carbon bond distances can be reduced as much as 5%, indicating a reorientation of many crystallites. The combined effect of these changes is a twofold to fivefold increase in electrical and thermal conductivity.

VIII. DEVELOPMENT OF NEW GRADES OF MANUFACTURED CARBONS

As these six examples indicate, manufactured carbon is, in reality, a family of versatile, self-lubricating materials. Throughout its early history and even into the mid-1950s, carbon manufacture was a pragmatic industry whose modest development effort was largely aimed at satisfying the chemical (chlor-alkali) and metallurgical industries. The advent of the nuclear and aerospace industries changed this. Where private industry could not justify the development costs to serve this exacting and unpredictable group of needs, governments funded developmental programs. This funding produced rapid and extensive disclosure of technology in this area. In a 1967 review, I cited over 100 references[4] on manufactured carbon for mechanical applications alone.

A predictable result of this injection of advanced technology is that the coming decade will see carbon products in greater variety and better quality than ever before. Even today each of the major manufacturers offers scores of different carbon materials. Selection of the best carbon for the application at hand has already become an engineering problem of some complexity.

IX. CARBON INDUSTRY TERMINOLOGY

Before there can be efficient and satisfactory communication between the carbon user and the carbon manufacturer, there must be a mutually understood language. In the carbon industry, there are a number of specialized terms.[5] A glossary, covering commonly used terms, appears in Appendix A.

X. MANUFACTURED CARBONS — TEST METHODS

Certain testing methods, like terminology, have become widely accepted in the manufactured carbon industry. Table 3.2 lists these. A detailed description of each will be found in Appendix B.

TABLE 3.2

Test Methods for Manufactured Carbon Articles

Property	Test method	Remarks	Ref.
Apparent density	ASTM C-559	Weigh part; calculate volume	5
	ANSI C-64.1 para. 4.8	Similar to C-559 but less precise	6
	ANSI C-64.1 para. 4.9	Weigh part in air and submerged in water; volume calculated by difference in weight	6
Ash content	ASTM C-561	Not suitable for metal-impregnated carbons	5
	ANSI C-64.1 para. 4.14	Can be used on silver- and copper-impregnated grades	6
Chemical analysis	ASTM C-560	For graphite and high purity carbons	5
Electrical resistivity	ASTM C-611	Requires rod or prism with length to diameter ratio of at least 6	5
	ANSI C-64.1 para. 4.4	Suitable for smaller specimens	6
Hardness — penetration	ASTM C-748	Suitable only on fine-grain carbons; uses Rockwell L with 0.250-in. ball	5
Hardness — rebound	ASTM C-886	Rebound height of a diamond-tipped hammer	5
Modulus of elasticity	ASTM C-647	Calculated from fundamental frequency of a vibrating rod	5
	ASTM C-769	Calculated from speed of sound as it passes through test specimen	5
Moisture content	ASTM C-562	Dries ground sample (60 mesh) at 105°C	5
Permeability	—	Test method being developed (see text)	—
Poisson's ratio	ASTM C-749	Generally 0.10 to 0.15 for manufactured carbon	5
Pore diameter	—	Force mercury into carbon pores (see text)	8
Porosity — open	ASTM C-373	Weight gain when impregnated with water	5
	ANSI C-64.1 para. 4.10	Weight gain when impregnated with SAE-10 oil	6
	—	Helium gas displacement, others	7
Porosity — total	—	Vol % = (2.26 — apparent density)/2.26 if ash content is low	—
Strength — compressive	ASTM C-695	Crushes cylinder with length to diameter ratio between 1.9 and 2.1	5
Strength — flexural	ASTM C-651	Four-point loading of a bar or rod	5
	ANSI C-64.1 para. 4.3	Three-point loading; gives higher values	6
Strength — tensile	ASTM C-565	Gage section of test specimen has a continuous radius	5
	ASTM C-749	Cylindrical gage section permits calculation of modulus and Poisson's ratio	5
Temperature limit	—	No established method (see text)	—
Thermal conductivity	ASTM C-714	Calculated from measurement of thermal diffusivity of a disk	5
Thermal expansion coefficient	ASTM E-228	Uses a quartz tube dilatometer	5

A. Permeability

For many applications, such as seals, the rate of fluid flow through the carbon is important. The darcy (or microdarcy) is the most common unit used to characterize permeability of porous solids. The manufactured carbon industry has not yet adopted a standarized test method for measurement of permeability.

Measurement of carbon permeabilities below 100 μD is difficult and time consuming. Hence, as a practical matter, seal carbon rings sold as "impervious" as pressure tested. They are clamped between plates, the internal volume is pressured with nitrogen gas (or dry air), and the exterior surface is painted with a soap solution to detect leaks. A seal ring sold as impervious to 0.69 MN/m^2 (100 psi) will show no bubbles after being pressurized for 1 min with dry nitrogen at this pressure.

B. Pore Diameter

The pore size distribution in carbon is usually measured by mercury porosimetry according to procedures first described by Washburn.[8] Mercury does not wet carbon; hence, pressure is required to force it into the pores. The relation between the pressure required and the pore diameter is set forth by the Kelvin equation:

$$d = (-4\,\delta \cos\theta)/p$$

where d = pore diameter (metres); δ = surface tension of mercury (0.48 N/m); θ = contact angle between mercury and carbon (130 to 140°); and p = pressure required (N/m^2). For mercury on carbon this reduces to:

$$d = (2.1 \times 10^2)/p$$

where d = pore entrance diameter in microns (10^{-6} m) and p = intrusion pressure, psi, or

$$d' = (1.4 \times 10^4)/p^1$$

where d′ = pore entrance diameter in angstroms (10^{-10} m) and p^1 = intrusion pressure (MN/m^2).

Figure 3.1 shows the pore size distribution plots for the carbon grades of Table 3.1. These plots were derived from the relationship between mercury penetration into the carbon and the pressure required to obtain this penetration. It assumes the pore entrances are round, which, of course, they are not. Microphotographs of carefully polished sections show that the pore cross sections of most carbon grades tend to be ellipses and crevices.

A second problem encountered in the determination of pore size spectrum is the fact that the high pressures needed to drive mercury into the smaller pores tend to collapse the structure. Dickinson and Shore[9] and Baker and Morris[10] found that the permanent alteration of the pore structure can occur at pressures as low as 19 MN/m^2 (2600 psi). This corresponds to the pressure required to intrude pore entrances of 0.08 μm (800 Å). The changes they observed were generally confined to pores smaller than 0.08 μm, and they were not significant until the pressure exceeded 35 MN/m^2 (4900 psi). Since the data for Figure 3.1 were obtained at pressures below 35 MN/m^2, the plots are reasonably indicative of pore structure of these six carbons.

"Median pore diameter" is that pore diameter at which half of the measured pore volume is in larger pores and half in smaller pores.

C. Temperature Limit

Carbon is oxidized by O_2 (air), CO_2, and H_2O. Of these, oxygen is by far the most

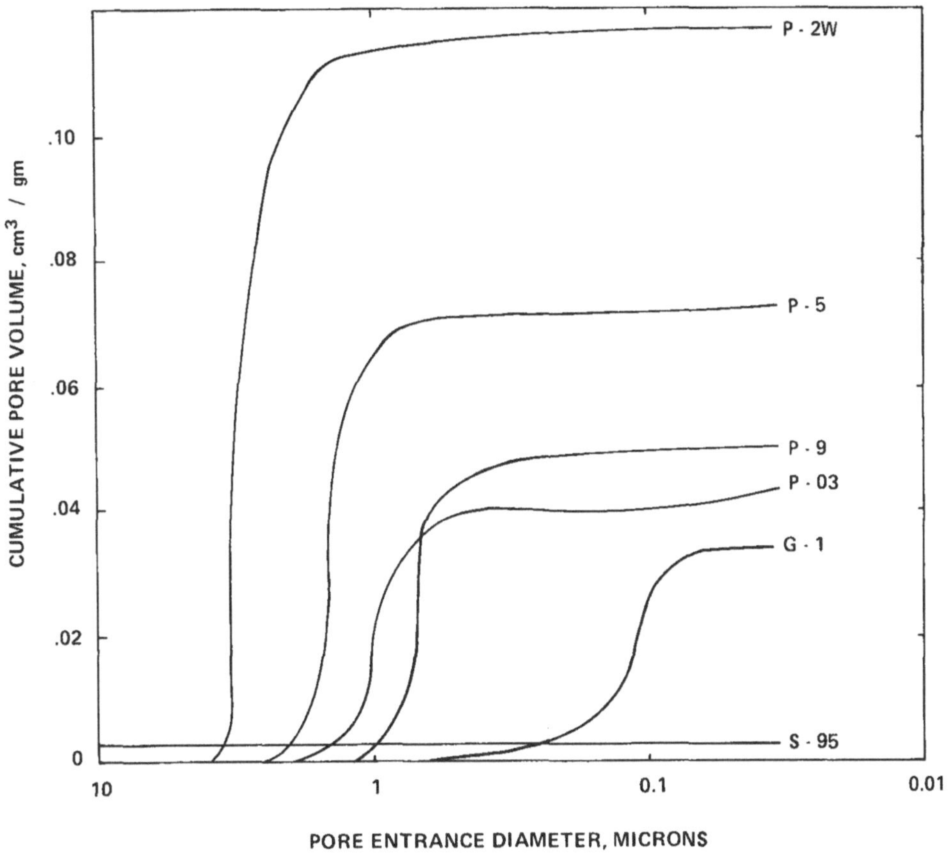

FIGURE 3.1. Pore size distribution.

corrosive, 1000-fold more reactive than CO_2 and 10,000-fold more reactive than H_2O. The rate of attack by oxygen is extremely sensitive to temperature, roughly doubling for each 28°C (50°F) increase. Moreover, carbons vary by at least five orders of magnitude in their reactivity toward oxygen. Most reactive are impure amorphous chars; least reactive toward oxygen are purified, large crystal-size graphites.

The "temperature limit" of nonimpregnated carbons is that temperature at which a 1-cm^3 carbon specimen loses weight at the rate of 0.1%/hr in moving air. The rate of weight loss generally increases with burn-off; hence, it is common to use the average rate between 0 and 5% burn-off. Temperature limit may be set by considerations other than oxidation rate. A resin-bonded carbon, for example, cannot be used at temperatures that soften or decompose the resin. Likewise, impregnated carbons cannot be used at temperatures which would cause the impregnant to decompose, to oxidize, or to melt and bleed out.

A third factor which sets a ceiling on the operating temperature of mechanical carbon is the requirement for dimensional stability. If a carbon is heated to a temperature well above the baking temperature used in its fabrication, it will shrink. The linear shrinkage can be as much as 1%, depending on the carbon type and the temperature involved.

Some manufacturers report two temperature limits for their carbons, a temperature limit in air and a temperature limit in a neutral atmosphere. The first is usually lower since it must consider chemical attack by oxygen.

XI. ANISOTROPY

Most carbon-graphites are anisotropic, that is to say, they have differing scalar properties depending upon the direction of measurements. The graphite crystal, it will be recalled, has a 100-fold variation in electrical resistance depending on the direction of measurement. To the extent that the fabrication processes line up these crystallites, the scalar properties of manufactured parts may be expected to be directional. Extrusion, for example, aligns acicular filler grains parallel to the extrusion axis. The crystallites within each grain tend to have their planes parallel to the major grain axis. This can produce with-grain and across-grain variations as high as 3 to 1 in properties such as electrical resistance and thermal conductivity.

Unidirectional compression molding produces less orientation, and isostatic compression molding least of all. A value of "negligible" anisotropy is commonly taken to mean that anisotropy, if present, produces effects less than one standard deviation of replicate measurements on the property in question.

XII. VARIATION IN PROPERTIES

As in any engineering material, the properties of a given grade of manufactured carbon will vary from specimen to specimen and from point to point within a specimen. The largest deviation occurs in scalar properties sensitive to flaws, for example, tensile strength or flexural strength. Nonscalar properties, such as apparent density, generally have less variation.

Shelton and Atkins[11] recently completed an intensive study on two 8-in. billets of medium grain-size graphite. They found the coefficient of variance, $s/\bar{x}$, for tensile strength measurements to be 12%. In this instance s was the standard deviation of replicate tensile strength measurements on bars cut from the billet. The average of all tensile measurements was designated $\bar{X}$.

A similar unpublished program conducted by ASTM while developing tensile test method C-565[5] found coefficients ranging from 7 to 16%, depending upon the type of manufactured carbon tested in tension. In general, lower coefficients were found in fine-grain, compression-molded impregnated carbon-graphites.

The above values are for specimens taken from a single lot of materials. Carbon users receive parts from many lots because their shipments extend over months or even years. To obtain information on lot to lot variation, a 1-year collection of quality control data for grade P-5 was examined. The test coupons, roughly 180 in number, were individually molded at a single pressure in a precision hydraulic press. They were then baked in a production furnace monitored by laboratory technicians. The number of coupons prepared each week ranged from 0 to 12 depending upon production requirements.

Figure 3.2 displays the values obtained for four physical properties. The abcissa on this plot is a probability scale. The data points lie in a line; hence, they have a "normal" distribution of values. Each curve is labeled with the property tested and its average value ($\bar{x}$) ± one sample standard deviation(s). Also shown is the coefficient of variation, $s/\bar{x}$. The coefficient of variation ranges from 0.96% for apparent density to 9.5% for flexural strength. I strongly suspect variations in the property values were caused by undetected lot to lot changes in the molding flour and by undetected changes in baking conditions over the year.

Table 3.3 summarizes the data of Figure 3.2. Production parts of grade P-5 (or any other carbon) are not generally prepared by laboratory personnel on precision hydraulic presses; hence, it is logical to expect a somewhat wider range in physical properties on them. Even so, the range is not so different from that seen on other engineer-

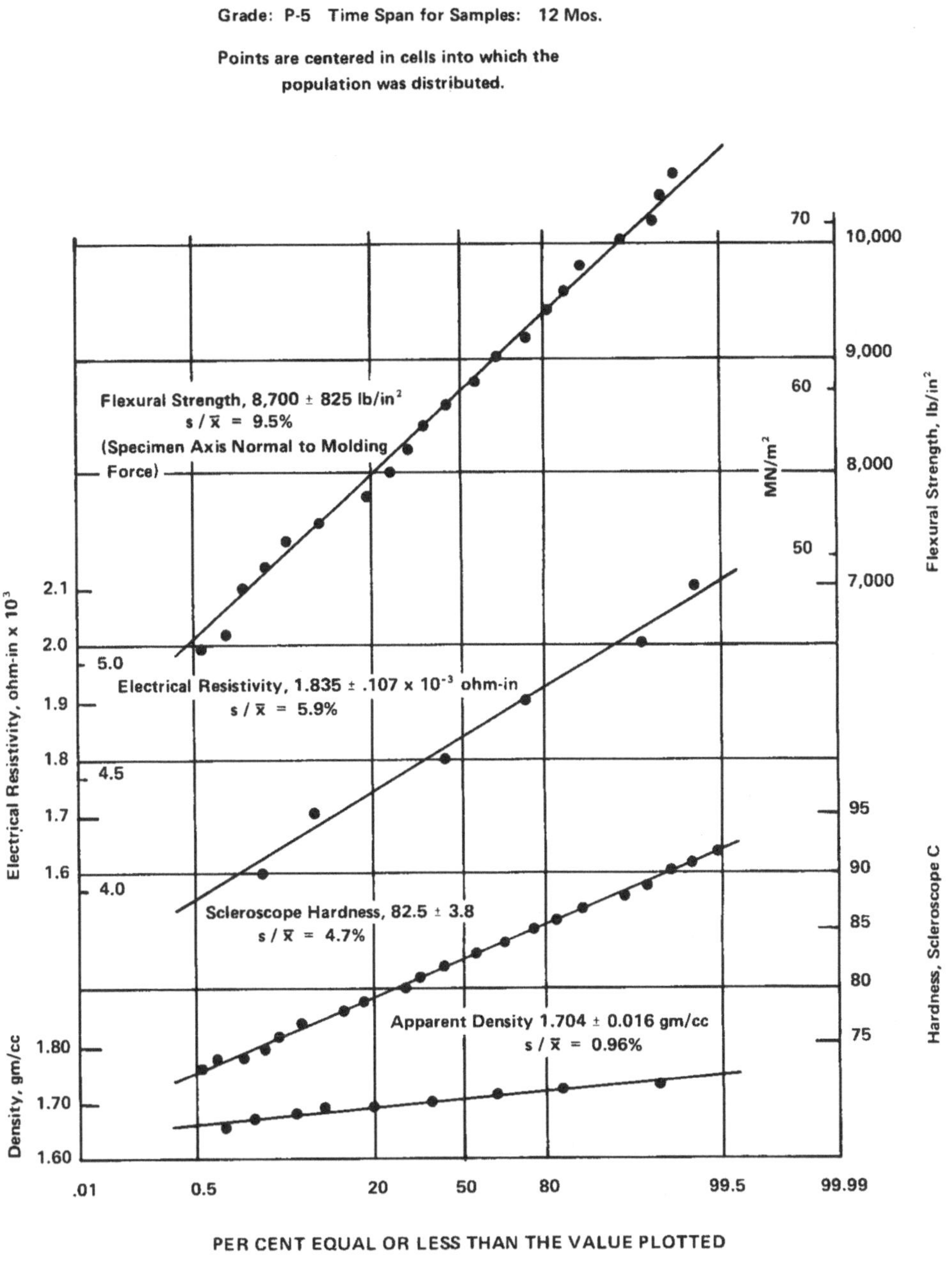

FIGURE 3.2. Range of physical properties.

ing materials. Lyman[12] reports the coefficient of variation for tensile strength of cold rolled 1035 steel bars in 6%; that for aircraft quality forged 7079-T6 aluminum is 3.3%.

Most carbon parts are impregnated with metals or resins to improve their properties. Low density, low strength parts take up more impregnant. Because of this, impregnated grades generally have a lower spread in physical properties than the unimpregnated base from which they are derived. Thus, impregnated grades should be used where a minimum spread in physical properties is required.

Another route to reducing the variation in physical properties is to have the entire production run made from a single lot of carbon. The blanks from which the parts

TABLE 3.3

Physical Properties — Minimum Range Obtainable on a Production Scale

	Average, $\bar{x}$	Standard deviation, s	Coefficient of variation, s/$\bar{x}$, %
Apparent density (g/cc)	1.704	0.016	0.96
Hardness (Scleroscope C)	82.5	3.8	4.7
Electrical resistivity			
Ω cm × 10^3	4.661	0.272	5.9
Ω in. × 10^3	1.835	0.107	5.9
Flexural strength			
MN/m^2	59.98	5.69	9.5
lb/in.^{2}	8700	825	9.5

Note: Grade, P-5; specimens tested, 180; time span for specimen manufacture, 12 months.

are machined should be molded rather than extruded, and they should be as near as possible to the finished size of the part. On long production runs where many lots are required, some reduction in the spread of physicals can be made by selecting molding flour blends and designating specific presses and furnaces to be used. Each such restriction increases cost to the manufacturer and generally results in a higher price.

REFERENCES

1. Beech, R. N. and Price, M. S. T., Copper graphite materials, *Industrial Carbons and Graphite*, Society of the Chemical Industry, London, 1958, 448.
2. Humenik, M. and Van Alsten, R. L., Metal Graphite Composites, U.S. Patent 2,983,034, 1961.
3. Humenik, M. and Van Alsten, R. L., Copper Graphite Composites, Canadian Patent 589,361, 1959.
4. Paxton, R. R., *Electrochem. Technol.*, 5, 174, 1967.
5. ASTM Standards — Part 17, American Society for Testing and Materials, Philadelphia, 1977.
6. *Brushes for Electrical Machines*, American National Standards Institute, New York, ANSI/NEMA No. C 64.1-1970, 1970, 32.
7. Bond, R. L., *Porous Carbon Solids*, Academic Press, New York, 1967.
8. Washburn, E. W., *Proc. Natl. Acad. Sci. U.S.A.*, 7, 5, 1921.
9. Dickinson, J. M. and Shore, J. W., *Carbon*, 6, 937, 1968.
10. Baker, D. J. and Morris, J. B., *Carbon*, 9, 687, 1971.
11. Shelton, W. R. and Atkins, G. R., The Correlation of Nondestructive Test Techniques with the Fracture Behavior of ATJ-S Graphite, AFML-TR-72-38, Air Force Materials Laboratory, Wright-Patterson Air Force Base, Ohio, August, 1972.
12. Lyman, T., Ed., *Metals Handbook*, Vol. 1, 8th ed., American Society for Metals, Metals Park, Ohio, 1961, 879.

Chapter 4

MANUFACTURE OF CARBON ARTICLES

I. INTRODUCTION

The carbon industry, like other branches of the chemical processing industry, is noticeably reticent when it comes to the details of its manufacturing processes. The reason for this reserve is not difficult to understand. Laboratory development costs for a new type (grade) of carbon material range from $5000 to $200,000. On top of this cost is the investment for special manufacturing facilities for the new grade plus the usual commercial costs of introducing a new product. Balanced against this outlay is the manufacturer's hope that the new product will outperform competitive products for a sufficient time to allow him to recoup his investment and make a profit. The odds of profiting from his investment are not helped by telling the competitor how to make this new product.

The novelty of a new carbon grade usually lies in its formulation and the processing steps, items hard to protect adequately by patent coverage. Rarely is the end product distinctive enough to merit a composition-of-matter patent. Most manufacturers, therefore, choose the course of secrecy for the majority of their developments.

The last decade, however, has brought a considerable opening up of the general features of carbon technology to public view. Research sponsored in the U.S. by the Atomic Energy Commission, the National Aeronautics and Space Administration, and the Department of Defense and overseas by similar European agencies has strengthened the scientific base of the carbon industry.[1,2] As carbon manufacture lost its aura of being a "black art", disclosures began to flow from the industrial practicioners as well. One of the most useful developments was the founding in 1961 of the technical journal *Carbon*. Drawing from these sources, plus the patent literature, one is able to put together a fairly clear picture of the manufacturing processes used to make carbon articles.

II. RAW MATERIALS

The raw materials used in carbon manufacture can be divided into three general groups: fillers, binders, and additives (Figure 4.1). Carbonaceous fillers make up 20 to 80% of the final item. Filler grains are held together with binder. Additives are used to modify the final product.

A. Fillers

Principal filler materials are calcined petroleum coke, natural graphite, synthetic graphite, and carbon black.

Petroleum coke is a by-product of the petroleum industry made by a batch process (delayed coking) or a continuous process (fluid coking). Petroleum coke is usually calcined at 900 to 1200°C (1625 to 2192°F) before grinding and sizing for the carbon industry. The source of the crude petroleum used, coking conditions, calcining temperature, and particle size (and shape) all affect the final manufactured carbon item.

Natural graphite is mined in Texas, Mexico, Korea, Austria, Madagascar, and Ceylon. Ash content, ash particle size, and ash composition are the principal variables among commercial natural graphites. Other factors of importance are porosity, size and shape of the graphite particles, and perfection of the laminar graphite crystals within each particle.

RAW MATERIALS

FILLER	(20 - 80%) NAT. GRAPHITE, SYN. GRAPHITE PETR. COKE, LAMPBLACK CHARCOAL
BINDER	(15 - 50%) SYNTHETIC RESINS COAL TAR PITCH PETROLEUM PITCH METALS CARBOHYDRATES
ADDITIVES	(0 -10%) ABRASIVES FILM FORMERS ANTIOXIDANTS GRAPHITIZING AIDS

FIGURE 4.1. Raw materials used in mechanical carbons.

Synthetic graphite may be calcined petroleum coke which has been heat-treated above 2500°C (4532°F) to enlarge the crystallites. Often as not, it is ground up scrap electrodes (electrographite) and, hence, includes graphitized pitch coke as well as impurities picked up by the electrode.

Lampblack is made by the pyrolysis of natural gas or the combustion of fuel oil in an oxygen-poor atmosphere. There are scores of grades of black designed for uses ranging from printing ink to rubber reinforcement.

B. Binders

The most common binder for carbon articles is coal tar pitch. This is a by-product of the steel industry (metallurgical coke production) and is widely available. Other binders include metals, petroleum pitch, thermosetting resins (commonly phenol-formaldehyde resins), and carbohydrates such as sugar or molasses.

C. Additives

Abrasives such as finely divided silica, silicon carbide, or tantalum/niobium carbide are often added to the carbon mix to control film formation at the rubbing carbon-metal interface.[6] Other additives such as molybdenum disulfide, barium fluoride, or silver sulfide are placed in the mix to enhance film formation. Johnson and Nickerson[3] describe the addition of organophosphates to the mix to enhance oxidation resistance of the baked carbon article.

III. FORMULATION

Obviously there are many possible raw materials for the formulator to draw upon. Considerable experience is required to select the appropriate materials and to combine them in just the right proportions. In making this selection, the formulator must consider both the cost of the ingredients and their effect on yield and properties of the final product. Skillful practitioners of this art are treated with the deference accorded the brewmaster in a successful brewery.

The first phase in the manufacture of molded carbon parts is preparation of the molding flour. Figure 4.2 outlines the major operations.

A. Mixing

A uniform distribution of binder, filler, and additive is essential to the production of quality carbon products. Large, heavy-duty, double-armed dough mixers jacketed so they can be heated to temperatures as high as 325°C (617°F) are the standard in the industry. Where mixing is done at temperatures below the melting point of the binder, volatile solvents such as benzene, xylene, ethylene dichloride, alcohol, or acetone are used to help disperse the binder. Mixing times range from 1 to 10 hr, depending upon the temperatures used and the particle size range of the filler. Figure 4.3 is an illustration of a typical mixer. If the product is to be formed by extrusion, hot mix is transferred directly to the extruder.

Metal-bonded mixes are usually prepared by cold blending the metal and graphite powders. Considerable care is required to obtain and keep a uniform blend of dense metal particles throughout the mix.

B. Molding Flour Preparation

Compression molding is the usual way of forming carbon for mechanical components. To prepare a suitable molding flour, the hot plastic lumps from the mixer must be extensively processed (Figure 4.2).

The lumps from the mixer are dumped into large cooling trays. When cool, they are reduced to a pea-size grist in a jaw crusher. Then they are milled in a hammer mill to yield a fine flour, usually 90% through 200 mesh. The fine flour from several mixes of a given formulation is commonly blended together to smooth out minor mix to mix variations. Trace amounts of die lubricants such as stearic acid are added during this blending operation.

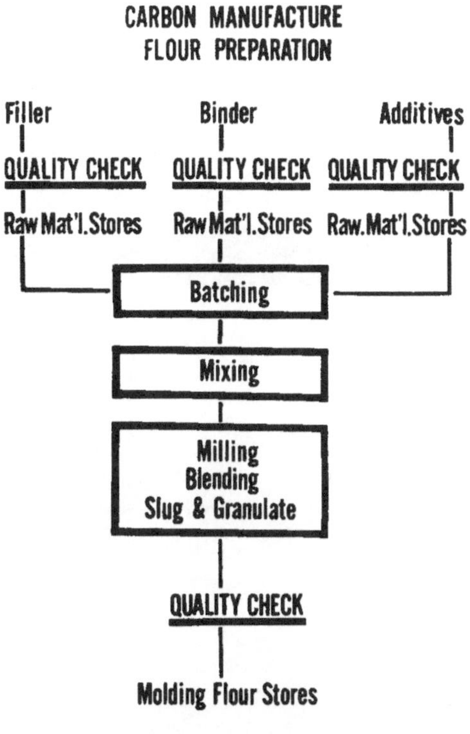

FIGURE 4.2. Carbon manufacture — flour preparation.

FIGURE 4.3. Typical heavy-duty double-armed dough mixer.

The low bulk density of blended flour, coupled with its poor bulk flow characteristics, renders it unsuitable for most molding operations. The usual practice, therefore, is to press this flour into low density preforms, granulate the preforms, and screen the resulting powder. Coarse particles are recycled to the granulator while fine particles are returned to the preform press.

Completion of these operations ends the first major step in the formation of manufactured carbon parts.

IV. PRODUCING CARBON BLANKS

The second major step in the manufacture of a carbon component is preparing the blank. Figure 4.4 outlines the major process steps.

A. Compression Molding

Molding flour, prepared as described above, is placed in hardened steel dies and pressed. Pressing forces range from 28 to 280 MN/m^2 (2 to 20 ton/in.^{2}). Plates and disks are the usual shapes. Thickness is normally kept below 50% of the width (or diameter) to avoid density gradients due to die wall friction. Area of the plate (or disk)

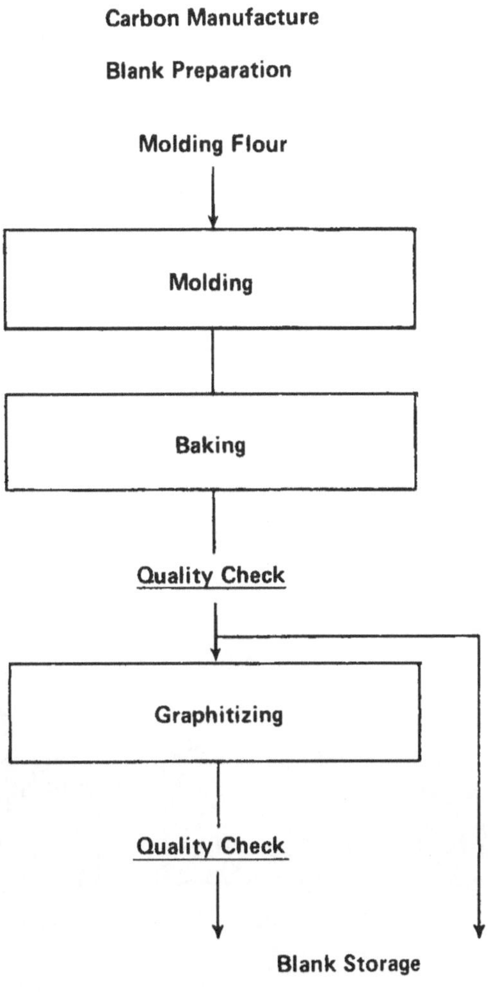

FIGURE 4.4. Carbon manufacture — blank preparation.

is limited by the press capacity. A 0.30 × 0.30 m (12 × 12 in.) plate, for example, would require a 13.4-MN (1500 t) clamping force to mold a flour requiring 190 MN/m² (10 t/in.²) for proper compaction. Figure 4.5 shows a 0.5-MN (60 t), fast-acting hydraulic press.

Modern presses utilize core pins so that rings as well as disks can be formed. A contoured die cavity permits the molding of parts with complex external shapes incorporating such features as flats, notches, or ears on the outer diameter. Contoured faces on the top and bottom punch permit the molding of features on these surfaces as well. Figure 4.6 shows a typical die and punch set. The relationship between part design and die design is so important a topic that a later section will be devoted to a discussion of the problem.

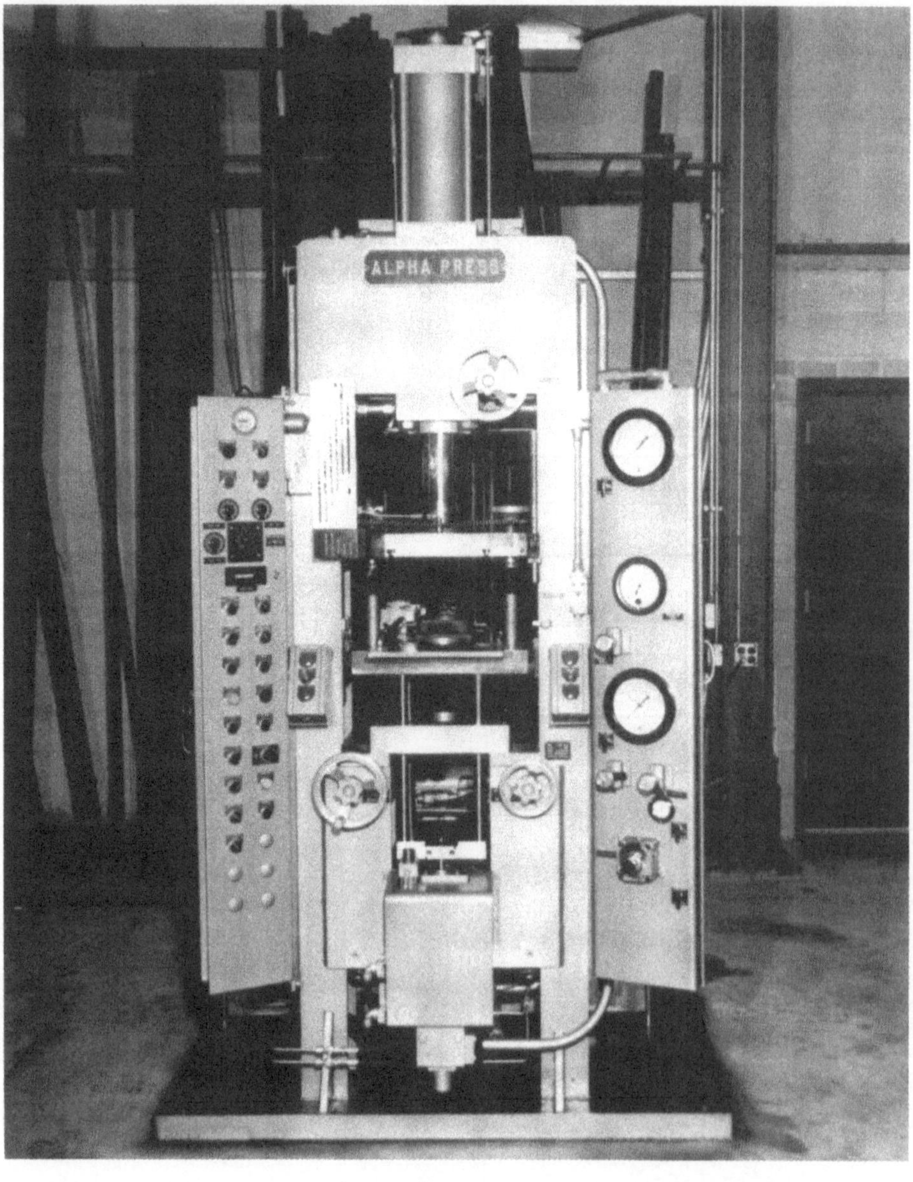

FIGURE 4.5. Fast-acting hydraulic press.

FIGURE 4.6. Carbon manufacture — typical die set.

B. Extrusion

The cheapest method of forming carbon is by extrusion. The hot mix, usually plasticized by a small amount of oil, is dumped into a heavy-walled cylindrical chamber. One end of this chamber contains a forming nozzle. The plastic mix is forced through this nozzle by a hydraulic powered ram. Cylinders up to 1 m (39 in.) in diameter are extruded, as are slabs of various width to thickness ratios. The extruded shape is cut into 1.3- to 2.3-m (4- to 8-ft) lengths for handling, and these are chilled in a water bath to harden them. Extruded carbons generally contain larger grains and larger voids than molded carbons and have lower strength and density. In addition, extruding tends to give an anisotropic part with differing physical properties when the part is measured parallel to, and normal to, the axis of extrusion. Where very large parts are required, however, extrusion was, until recently, the only practical forming method.

C. Isostatic Pressing

A new forming process, isostatic pressing, represents a compromise between extrusion and molding. In this process, the finely divided molding flour is placed in a shaped, sealed rubber bag. The bag is submerged in liquid in a pressure vessel; the vessel is then closed and the pressure raised. Pressure requirements are somewhat less than for compression molding because there is no die friction. Graphite articles as large as 0.40 m (16 in.) in diameter × 1.8 m (72 in.) long formed by isostatic pressing are now commercially available.

The quality of these parts is every bit as good as that of much smaller compression-molded shapes of the same flour. Figure 4.7 shows the interrupted thread closure and top of an isostatic press capable of accepting bags up to 0.6 m (24 in.) in diameter × 3 m (118 in.) long. Standing beside the unit is a typical isostatically pressed part.

V. BAKING

After carbon shapes are formed, they must be baked to convert the organic binder into carbon. This is the most critical step in the entire manufacturing process. Both the atmosphere around the parts and the time-temperature program must be controlled if sound, high-strength parts are to be obtained.

During the early stages of baking, i.e., below 425°C (797°F) the parts soften and

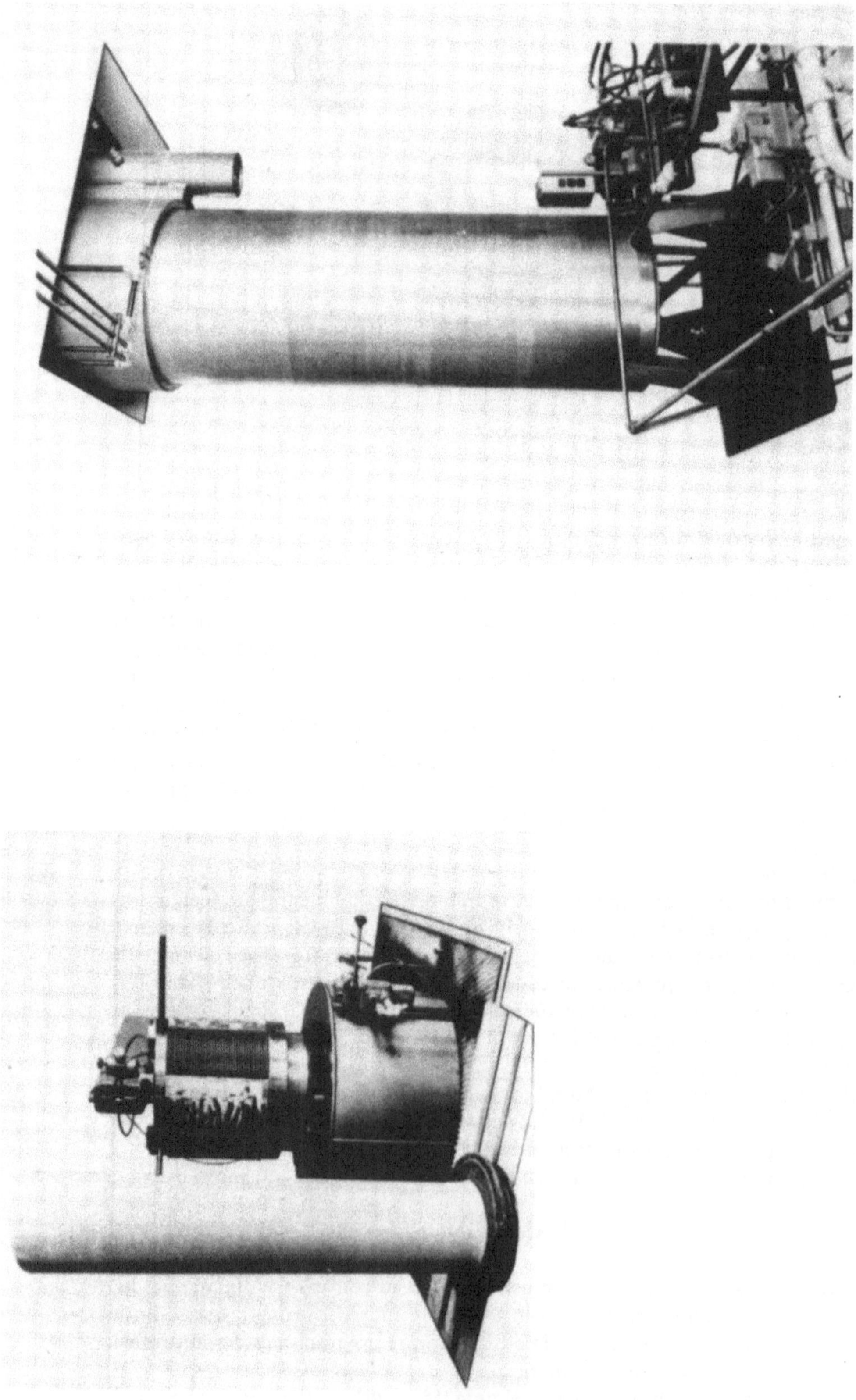

FIGURE 4.7. Isostatic press.

expand. Later they shrink, and the final volume may be as little as 65% of the starting volume. Each part must be supported during baking so that these changes can be accommodated without applying forces that will distort or rupture the part. This support is commonly supplied by bedding the parts in a granular mixture of graphite, coke, charcoal, and sand. The baking of rings presents a special problem because a shrinking inner diameter closing in on the incompressible bedding mixture could generate hoop stresses high enough to rupture the ring. To alleviate this, properly sized "shrink sticks" of cardboard or wood are placed inside each ring displacing part of the bedding. The "shrink stick" carbonizes and shrinks at the same rate as the ring inner diameter, and destructive hoop stresses are avoided.

Considerable quantities of gas are evolved as the binder is pyrolyzed. This gas, generated throughout the body of each part, must escape. If parts are baked too rapidly, or if they are pressed to too high a density, internal gas pressure will rupture the part. Occasionally, these laminar voids do not extend to the surface of the part, and the defect goes undiscovered until the part is machined. In 99% of the cases, however, overstressed parts come out of the bake in pieces. The only defense against this loss is exacting control of the molding and baking steps. Baking cycles run as long as 8 weeks for large, high-density parts. During certain parts of the cycle, the rate of temperature rise is less than 0.5°C/hr (1°F/hr). Smaller parts can, of course, be baked more rapidly.

Peak temperatures for carbonizing bakes range from 700 to 1300°C (1292 to 2372°F) depending upon the properties that are desired. In general, the higher the baking temperature, the harder, more abrasion resistant is the part. High baking temperatures also impart chemical resistance and dimensional stability to the carbon part and increase its strength and thermal conductivity.

The baking of carbon parts is done in kilns so designed that both the temperature and the composition of the gases surrounding the parts can be closely controlled. Small concentrations of oxygen in the baking atmosphere, for example, drastically reduce both the shrinkage and the strength of the baked part. Figure 4.8 shows operators unloading a typical batch kiln.

FIGURE 4.8. Unloading a typical batch kiln.

VI. GRAPHITIZING

Graphitizing greatly increases the resistance of the part to thermal shock and chemical corrosion by oxygen (air). In this process, parts are heat-treated to a temperature between 2475 and 3025°C (4487 to 5477°F). Residence time at the peak temperature need not be long; indeed, on continuous graphitizers the entire cycle from room temperature to 2800°C (5072°F) back to room temperature takes less than 3 hr. Large shapes require the gentler heating of batch graphitizers, where the cycle is normally 2 weeks or more. Figure 4.9 shows the charging port of a continuous graphitizer. Figure 4.10 shows three batch graphitizers. The one in the foreground is loaded and being heated, the middle one is being loaded, and the back one is being disassembled after completion of a graphitizing cycle. Figure 4.11 shows a batch graphitizer being unloaded after it has cooled.

VII. HOT PRESSING

Where maximum density is required, parts are graphitized under a compressive load. Deformation in the order of 50% is obtained, and the typical increase in density is from 1650 to 1950 kg/m^3 (1.65 to 1.95 g/cc). The resulting part is highly anisotropic; electrical resistance, for example, is ten times higher in the direction of compression than in the lateral direction. This anisotropy can be a distinct virtue which permits the design of a part with truly unique characteristics. As an example, susceptors for induction furnaces can be built in which the heated zone can be confined to a narrow portion of their axial length.

VIII. CURING RESIN-BONDED CARBON

Most resin-bonded carbon parts are cured under pressure in the forming press. Post cures to further crosslink and harden the resin are common, however. These are generally done in air-filled ovens using a controlled time-temperature cycle peaking in the 200 to 250°C (392 to 482°F) range. Post cure cycles are generally less than 16 hr long.

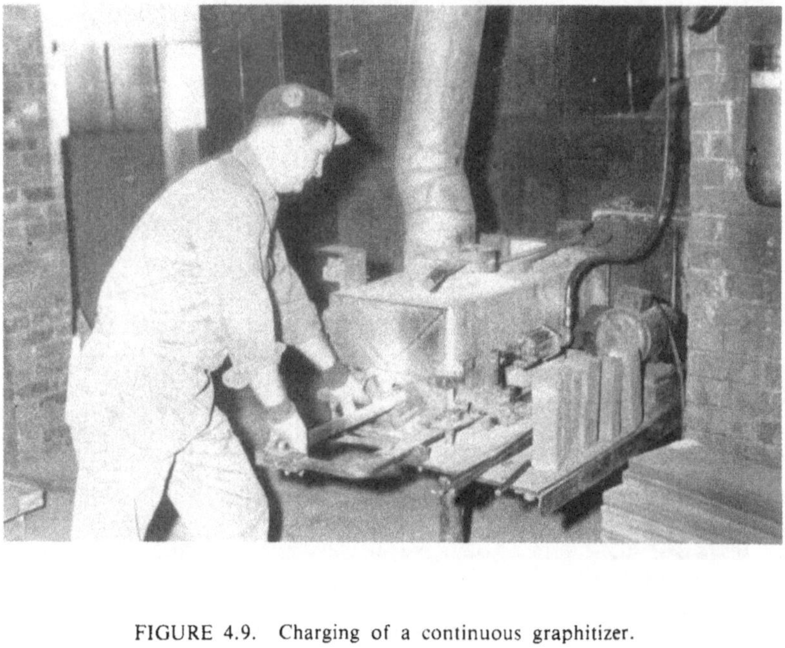

FIGURE 4.9. Charging of a continuous graphitizer.

FIGURE 4.10. Overall view of a bank of batch graphitizers.

FIGURE 4.11. Unloading a batch graphitizer.

IX. CURING METAL-BONDED CARBON

Metal-bonded carbon parts require essentially the same sintering conditions as would be required for the pure metal part. The presence of graphite permits a somewhat faster and higher temperature bake. Where carbide formation can occur, time at temperature must be kept to the minimum needed to secure a strong part. In some cases, induction heating followed by hot pressing or forging may be used to minimize carbide formation and retain some ductility.

X. FINISHING OPERATIONS

Figure 4.12 outlines the various operations involved in converting baked carbon blanks into manufactured parts ready for use in mechanical devices.

A. Impregnation

Carbon is rarely at its best in the as-baked or as-graphitized state. It contains a network of pores through which pyrolysis gases escaped during the baking process. This open porosity, measured by water impregnation, ranges from 3 to 30% by volume. Permeability ranges from 10 to 50,000 μD.[4] Gases and even liquids can pass through the more permeable bodies. This renders them unfit for use as seals. To reduce undesirable permeability, the carbon parts are impregnated.[5]

1. Thermoplastic Impregnants

Any impregnant which can melt and run back out of the carbon is considered a thermoplastic impregnant. Included in this class are waxes, thermoplastic resins, metals, inorganic salts, and ceramics. The melting point of the impregnant places an upper limit on the operating temperature of these impregnated carbon parts. This temperature limit may be lower than the temperature limit of the carbon base.

Table 4.1 shows the effect that four typical thermoplastic impregnants have upon the properties of base carbon grade P-5. A polyethylene impregnation converts the base into a substantially impermeable carbon which is suitable for use in mechanical seals. The low melting point of polyethylene imposes a temperature limit of 93°C (200°F) on P-5E. Accordingly, this grade is limited to low temperature applications such as dishwashers or oil seals which are lightly loaded. Grade P-5E could not be

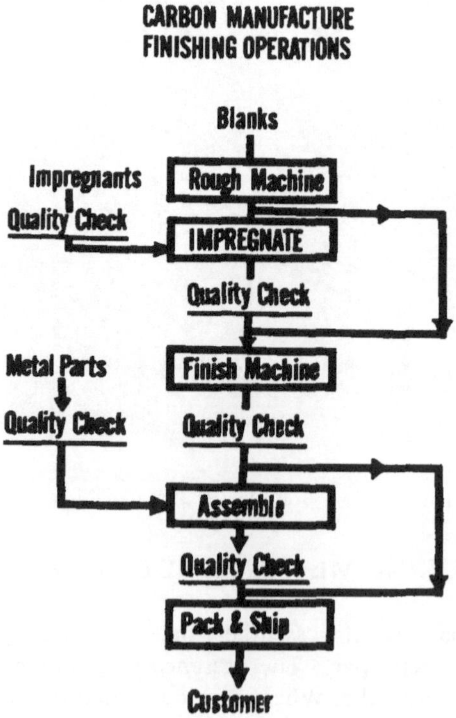

FIGURE 4.12. Carbon manufacture — finishing operations.

TABLE 4.1

Thermoplastic Impregnations for Mechanical Carbons

Identification[a]	P-5	P-5E	P-15	P-5Ag	P-5N
Impregnant	None	Polyethylene	Babbitt metal	Silver metal	Lithium fluoride
Density (g/cc)	1.70	1.80	2.40	2.35	1.85
Open porosity (vol %)	18	1	7	5	10
Permeability (μD)	6000	<0.1	1	100	100
Temperature limit in air °C(°F)	260 (500)	93 (200)	204 (400)	260 (500)	260 (500)
Hardness (Shore Scleroscope)	82	80	85	90	90
Strength MN/m² (psi)					
Compressive	172 (25,000)	172 (25,000)	258 (37,500)	310 (45,000)	241 (35,000)
Flexural	60 (8,700)	62 (9,000)	83 (12,000)	90 (13,000)	69 (10,000)
Tensile	41 (6,000)	48 (7,000)	62 (9,000)	69 (10,000)	48 (7,000)
Modulus of elasticity					
MN/m²	18×10^3	21×10^3	26×10^3	26×10^3	21×10^3
psi	2.6×10^6	3.0×10^6	3.8×10^6	3.8×10^6	3.0×10^6
Coefficient of thermal expansion					
mm/mm°C, 20 to temp limit	4.0×10^{-6}	4.3×10^{-6}	4.7×10^{-6}	4.7×10^{-6}	4.9×10^{-6}
in./in.°F, 68 to temp limit	2.2×10^{-6}	2.4×10^{-6}	2.6×10^{-6}	2.6×10^{-6}	2.7×10^{-6}
Thermal conductivity					
W/m°C	9	9	10	26	10
Btu/hr °F ft	5	5	6	15	6
Electrical resistivity					
Ω cm	4.6×10^{-3}	4.6×10^{-3}	4.1×10^{-3}	2×10^{-3}	4.6×10^{-3}
Ω in.	1.8×10^{-3}	1.8×10^{-3}	1.6×10^{-3}	0.8×10^{-3}	1.8×10^{-3}

[a] Material designations are those of Pure Carbon Company, Inc., St. Marys, Pa.

safely used, for example, in an automobile water pump or a water circulator for a home heating system because operating temperatures in these devices are often well above 93°C (200°F).

For higher temperature capability as well as substantially improved physical properties, babbitt metal may be impregnated into the pores of a carbon body. Grade P-15 (babbitt-impregnated P-5) makes an excellent seal nosepiece for high-pressure hot water circulating pumps.

Where a seal sees high face loads coupled with high rubbing speeds, the ability to conduct heat away from the interface becomes important. Impregnation of P-5 base with silver metal produces grade P-5Ag, a grade widely used for this type of duty. Silver metal does not wet and bond to carbon; hence, when it freezes in the pores and contracts, it leaves a residual permeability. This permeability is low enough to permit the use of P-5Ag in seals handling liquids, but it is too high to be acceptable in seals handling gases. Grade P-5Ag also finds wide use in bearings because of its high compressive strength and thermal conductivity.

The three impregnants described so far were chosen for their ability to decrease the permeability of carbon and strengthen its structure. In these grades, the carbon body itself normally generates the transfer film so necessary for a low friction coefficient and low wear rate. Carbon, however, cannot generate a transfer film in the absence of polar fluids, a situation that is found in vacuum systems, dry gas systems, and most cryogenic applications. These applications require the aid of a supplementary film former such as the lithium fluoride used in P-5N. In addition to its value as a supplementary film former, lithium fluoride is substantially insoluble in water, thermodynamically stable, and has a high melting point.

2. Thermosetting Impregnants

An unfortunate attribute of thermoplastic impregnants is their tendency to melt and bleed out of the carbon part if it is overheated. This bleeding can be avoided by using thermosetting resins, principally phenolics, epoxides, polyesters, and furanes. The trick here is to find resin systems which cure with minimum shrinkage, have low viscosity for easy impregnation, and have long pot life for manufacturing convenience. A further requirement is the need of conveniently removing excess resin adhering to the surface of the impregnated carbon.

Table 4.2 shows the improvement in physical properties obtained by impregnation with thermosetting resins. The upper limit of operating temperature for these carbons is the temperature at which the resin undergoes pyrolysis. Grade P-9J is too permeable for use in most seal applications. Because of its excellent film-forming characteristics, however, it finds wide use as piston rings in nonlubricated compressors. Grades P-89 and P-29 are widely used in seals with P-89 preferred where caustic chemicals are encountered.

3. Inorganic Salt Impregnants

The oxidation rate of manufactured carbon is controlled by its internal (micropore) surface area and by the amount and catalytic activity of the ash. Internal surface area can be reduced by using synthetic or natural graphite instead of charcoal fillers. Ash content can be minimized in a similar manner. As a practical matter, however, it is difficult to get ash content below 1% through formulation, and 1% ash can have a potent catalytic effect on oxidation rate.

Both ash content and internal surface area can be reduced by heat-treating the carbon parts at 2700 to 3200°C. If the parts are flooded with chlorine or hydrogen fluoride during heat treatment, ash content can be brought below 1 ppm (0.0001%). An alternative purification agent is Freon gas. This is easier to handle, and it breaks down to chlorine, fluorine, HCl, and HF at the heat-treating temperatures.

TABLE 4.2

Thermosetting Impregnations for Mechanical Carbons

Identification[a]	P-9	P-9J	P-89	P-29
Impregnant	None	Phenolic	Furane	Polyester
Density (g/cc)	1.75	1.80	1.85	1.85
Open porosity (vol %)	16	7	3	1
Permeability (μD)	1,000	300	0.5	<0.1
Temperature limit in air °C (°F)	260 (500)	260 (500)	260 (500)	232 (450)
Hardness (Shore Scleroscope)	60	65	80	75
Strength MN/m^2 (psi)				
Compressive	138 (20,000)	155 (22,500)	172 (25,000)	190 (27,500)
Flexural	48 (7,000)	55 (8,000)	69 (10,000)	72 (10,500)
Tensile	31 (4,500)	38 (5,500)	52 (7,500)	55 (8,000)
Modulus of elasticity				
MN/m^2	16×10^3	21×10^3	21×10^3	23×10^3
$lb/in.^2$	2.4×10^6	3.1×10^6	3.1×10^6	3.3×10^6
Coefficient of Thermal Expansion				
mm/mm°C, 20°C to temp limit	3.6×10^{-6}	5.2×10^{-6}	6.1×10^{-6}	5.4×10^{-6}
in./in.°F, 68°F to temp limit	2.0×10^{-6}	2.9×10^{-6}	3.4×10^{-6}	3.0×10^{-6}
Thermal Conductivity				
W/m °C	12	12	12	12
Btu/hr °F ft	7	7	7	7
Electrical resistivity				
Ω cm	3.8×10^{-3}	3.8×10^{-3}	3.8×10^{-3}	3.8×10^{-3}
Ω in.	1.5×10^{-3}	1.5×10^{-3}	1.5×10^{-3}	1.5×10^{-3}

Material designations are those of Pure Carbon Co., Inc., St. Marys, Pa.

Purification can be counted on to reduce the oxidation rate of most manufactured carbons to 1% or less of their unpurified precursor. This treatment, however, is expensive. It will also reduce the self-lubricating characteristics of most carbons by reducing their ability to chemisorb water vapor (and other aids) which are essential to the self-lubricating properties of carbon. Lastly, the removal of ash often results in the loss of a polishing agent needed to control thickness of the transfer film on the rubbing metal interface.

An alternative to removing the catalytic ash is to deactivate it by converting it to a form which has reduced catalytic activity. An effective catalyst poison is phosphorous pentoxide. Experiments have shown that the addition of 10 ppm (by volume) of P_2O_5 to the air surrounding the carbon part reduces oxidation rate of the carbon by a factor of 10 to 100.

It is generally impractical to add P_2O_5 to the air around the part. Instead, one fills the pores of the carbon part with a compound which generates P_2O_5 at elevated temperatures. Careful choice of impregnant can also enhance film formation on the mating face at high temperatures (or low humidities) and thereby improve the self-lubricating properties of the carbon part.

Published research on this type of impregnant was summarized by Paxton.[6] The most effective salts appear to be acidic phosphates of amphoteric metals, particularly aluminum and zinc.[7,8] A water solution of the salt is forced into the pores of the carbon, the water is evaporated, and then the carbon is baked to convert the dried salt into a less soluble form. The quantity of salt left in the carbon after each impregnating cycle is small, and there is generally little effect upon room temperature physical properties. Grade P-03XHT, shown in Table 4.3, is an example of this type of impregnation.

A second type of impregnation forces molten phosphate salts[9] into the pores of the

TABLE 4.3

Inorganic Salt Impregnations for Mechanical Carbons

Identification*	P-03	P-03XHT	P-2003
Impregnant	None	Inorganic	Inorganic
Density (g/cc)	1.80	1.91	2.00
Open porosity (vol %)	8	5	0.3
Permeability (μD)	1,000	400	<0.1
Temperature limit in air [°C (°F)]	538 (1,000)	650 (1,202)	680 (1,256)
Hardness (Shore Scleroscope)	75	75	80
Strength [MN/m^2 (psi)]			
Compressive	138 (20,000)	155 (22,500)	190 (27,500)
Flexural	52 (7,500)	55 (8,000)	76 (11,000)
Tensile	34 (5,000)	41 (6,000)	52 (7,500)
Modulus of elasticity			
MN/m^2	14×10^3	14×10^3	17×10^3
$lb/in.^2$	2.0×10^6	2.0×10^6	2.4×10^6
Coefficient of thermal expansion			
mm/mm°C, 20°C to temp limit	3.6×10^{-6}	3.8×10^{-6}	4.5×10^{-6}
in./in.°F, 68°F to temp limit	2.0×10^{-6}	2.1×10^{-6}	2.5×10^{-6}
Thermal conductivity			
W/m °C	69	69	69
Btu/hr °F ft	40	40	40
Electrical resistivity			
Ω cm	1.5×10^{-3}	1.5×10^{-3}	1.5×10^{-3}
Ω in.	0.6×10^{-3}	0.6×10^{-3}	0.6×10^{-3}

* Material designations are those of Pure Carbon Co., Inc., St. Marys, Pa.

carbon. This strengthens the carbon and greatly reduces its permeability. Reduced permeability, in turn, further increases oxidation resistance by slowing diffusion of oxygen into the interior of the part. Another advantage of the molten salt approach is that it permits impregnation with relatively insoluble glass-like phosphates. Grade P-2003 of Table 4.3 is an example of this type of impregnant.

Carbons impregnated with soluble salts must be handled with care to avoid exudation. These carbons tend to absorb water from humid air. Water slowly softens the additive, converting it into a viscous gel. If the part is then quickly heated, this gel may exude and harden into a surface glaze. Slow drying, on the other hand, tends to leave a flaky deposit on the surface.

These exuded deposits are not generally abrasive. They may temporarily raise the coefficient of friction, but they do not usually affect wear rates. They can, however, cement together close-fitting carbon parts or carbon to metal interfaces. This can render a segmented circumferential shaft seal inoperative. Furthermore, when moist, the exudate is usually acidic and can accelerate corrosion of many metals. When dissimilar metals are in contact with these carbons, galavanic action can aggrevate corrosion effects. For this reason, 300 series steels, high nickel alloys, or other corrosion-resistant materials are recommended for contact with this class of impregnated carbon grades.

4. Siliconized Carbons for Enhanced Chemical and Abrasion Resistance

At high temperatures, carbon reacts with silicon containing compounds to form a family of materials known as siliconized carbons. The surface layer of siliconized carbon items contains silicon carbide, carbon, and occasionally metallic silicon. The silicon carbide content of the surface layer ranges from 99.99% to as low as 50% in materials used for sliding contracts. This concentration is controlled by varying either the substrate carbon or the time and temperature used for the siliconizing steps.

High purity silicon carbide surfaces are produced by chemical vapor deposition (CVD) process. Surfaces relatively rich in carbon are produced by a chemical vapor reaction (CVR) process.

5. CVD Plating for Impervious Oxidation-Resistant Surfaces

In the CVD or vapor plating process, a vapor containing both silicon and carbon is passed over a heated graphite mandrel or substrate. Typical vapors are mixtures of silicon tetrachloride and acetone,[10] silicon tetrachloride, benzene and hydrogen,[11] or methyltrichlorosilane and hydrogen.[12-14] Graphite temperatures range from 1175 to 1775°C (2147 to 3227°F) with 1400°C (2552°F) being the most commonly recommended value. At this temperature, silicon carbide plating deposits at a rate of 0.4 mm/hr (0.016 in./hr) if an ample supply of vapor reactants is provided.

As plated parts become larger than a few millimeters in diameter, increasing attention must be directed to getting a close match between the thermal expansion of the graphite substrate and the silicon carbide coating. Further, the graphite must be isotropic, that is, have the same expansion characteristics in all three axes. If the thermal expansion characteristics of the substrate and coating are not closely matched, the coating will crack or the part will warp when it is cooled from the plating temperature. Sometimes this cracking or warping does not become apparent until after a number of heating and cooling cycles for the plated article have taken place.

As an aid to getting a good bond between the substrate and the coating, the graphite may first be treated with silicon.[15] At 1425°C (2597°F), metallic silicon wets carbon and diffuses through its porous structure reacting slowly. This modifies the outer few millimeters of the graphite surface for the subsequent vapor plating operation and is said to produce a more adherent plating.

Vapor-plated plates as large as 25 × 300 × 1200 mm (1 × 12 × 47 in.) are commercially available. Tubes up to 300 mm (12 in.) outer diameter × 1400 mm (55 in.) long coated on all surfaces can also be purchased in the U.S. These parts typically have a plating thickness of 1.5 mm (0.059 in.). To avoid stress concentrations, edges and corners usually have a generous radius, 0.5 mm (0.020 in.) or more.

Table 4.4 gives properties of a typical CVD-coated item. The value for any particular part must obviously depend on the ratio of plating thickness to total thicknesses and upon the type of graphite substrate used.

6. CVR-Treated Carbons for Abrasion Resistance and Low Wear Rates

Parts for mechanical items such as seals must be manufactured to close tolerances, typically 0.1 mm (0.005 in.) or better. Plated parts would require diamond grinding to achieve these kinds of tolerances because the thickness of plating cannot be closely controlled. This would be extremely expensive. Accordingly, siliconized carbon parts for mechanical applications are generally produced by the CVR process which permits close dimensional control. In addition, the CVR process produces a surface layer containing 10% (or more) graphite which greatly aids in lowering friction and wear rates at rubbing surfaces. Figure 4.13 illustrates the type of surface modification produced by the CVR process.

Potter and Tone laid the foundation for the CVR process at the turn of the century.[16-19] They discovered that silicon monoxide reacts with carbon bodies at high temperatures to form a silicon carbide body identical in shape and texture to that of its precursor. Their work lay dormant for almost 60 years; then as an outgrowth of the CVD process, it was exumed. In the present embodiment of this process, completely machined graphite parts are exposed to silicon monoxide vapor at 1625 to 2225°C (2957 to 4037°F). Silicon monoxide reacts with the carbon:

$$2C + SiO \longrightarrow SiC + CO$$

TABLE 4.4

Siliconized Carbon

Identification[a]	CVD PE-6707	CVR PE-6923
Impregnant	None	Organic
Density (g/cm^3)	2.2	2.15
Open porosity (vol %)	0.0	0.8
Temperature limit in air °C (°F)	1,725 (3,137)	260 (500)
Hardness		
Rockwell 15T	100	93
Moh's	9.5	9.5
Knoop [100g], kg/mm^2	3300	400
Strength, MN/m^2 (psi)		
Compressive	345 (50,000)	138 (20,000)
Flexural	138 (20,000)	55 (8,000)
Tensile	55 (8,000)	34 (5,000)
Modulus of elasticity MN/m^2 (psi)	69×10^3 (10×10^6)	28×10^3 (4×10^6)
Coefficient of thermal expansion, mm/mm°C (in./in.°F)	4.4×10^{-6} (2.4×10^{-6})	4.4×10^{-6} (2.4×10^{-6})
Thermal conductivity, W/m°C (Btu/hr ft °F)	172 (100)	69 (40)
Electrical resistivity, Ω cm (Ω in.)	4×10^3 (1.6×10^3)	1.9×10^{-3} (0.8×10^{-3})

[a] Material designations are those of Pure Carbon Co., Inc., St. Marys, Pa.

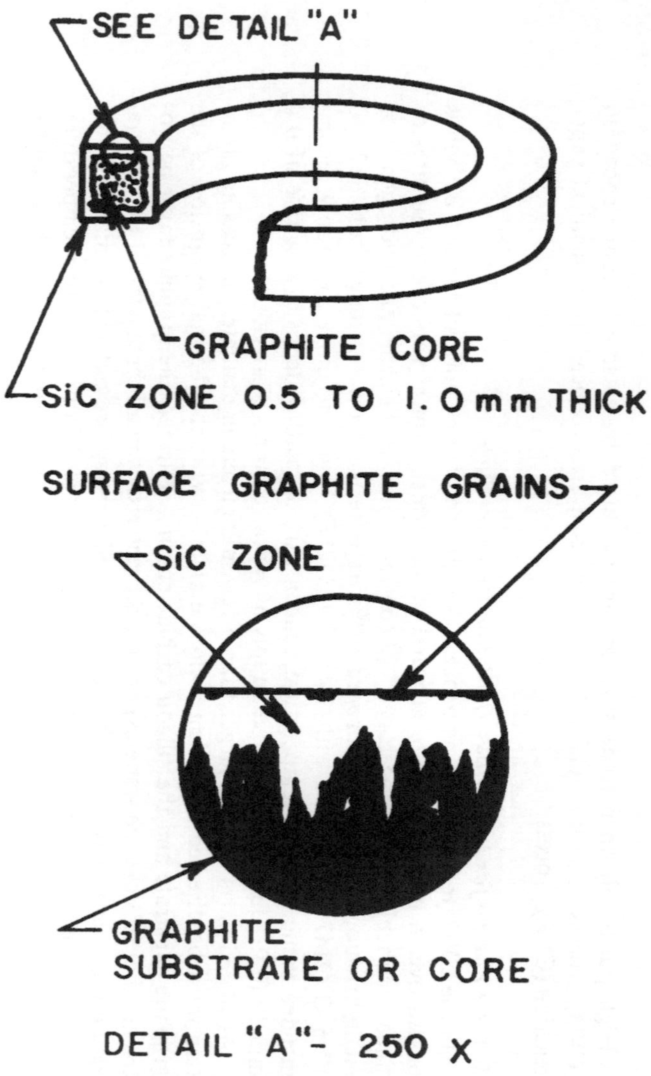

FIGURE 4.13. Siliconized carbon part prepared by the "CVR" process.

The SiC occupies essentially the same volume as two carbon atoms; hence, dimensions of the reacted part are essentially unchanged.

Silicon monoxide is generally produced by reduction of silica (SiO_2). Some workers[20-23] prefer silicon metal as the reducing agent; others[24] use carbon, or hydrogen.

Silicon monoxide vapor is then passed over the heated graphite part. The graphite surface temperature, along with the graphite type and reacting vapor composition, controls the concentration gradient of SiC in the finished part. A typical graphite requires 2 hr at 1925°C (3497°F) to produce a reaction layer 0.5 mm (0.020 in.) deep. Concentration of SiC in the surface layer increases as one approaches the exposed surface. Even there, however, 10 to 50% of the surface remains unreacted graphite. Islands of unreacted graphite on the surface can be directly observed under a microscope using polarized incident light. Knoop hardness tests show portions of the surface to be less than one tenth as hard as the majority of the area. Figure 4.14 pictures this surface.

Selection of the proper graphite substrate is second only to control of reaction condition in producing a quality part. Because the temperature is 500°F higher than in the CVD process, matching of thermal expansion characteristics is more critical — and more difficult. Other critical factors are porosity of the graphite, pore size distribution, and chemical reactivity of its filler and binder constituents.

Like all graphites, the CVR-treated graphite is porous and, hence, not suitable for applications such as seal rings without further processing. Impregnation with suitable resins, however, fills the pores and stiffens the part, rendering it valuable as a rubbing element in mechanical seals and other devices. Probably the most promising application is as a mating face for carbon seal noses. Properties of a typical seal ring/mating face material are shown in Table 4.4.

One of the particularly valuable attributes of CVR-treated graphites is their combination of extreme hardness with self-lubricating ability. Because of the free graphite on their surface, CVR-treated graphites can rub dry without seizing or galling. In addition, CVR-treated graphites have all the chemical resistance and thermal shock resistance of graphite. CVR-treated graphite will surely find increasing applications where wear due to corrosion or abrasion is a serious problem.

XI. SILICON CARBIDE BONDED SHAPES

An alternate to treating carbon substrates to produce surfaces rich in silicon carbide is to start with silicon carbide powder to produce a body that is mostly SiC. Such bodies are harder and stronger than carbon and can be used at higher temperatures.

Reaction bonded SiC starts with alpha SiC powder containing a small amount of graphite powders. A fugitive binder, usually a phenolic resin with a small amount of plasticizer, is added. The resulting mixture is pressed in hardened steel dies much like those used for carbon and then cured to produce a solid body. This resin-bonded body can be machined with diamond tipped tools to produce the desired shape. The shaped body is then carbonized and placed in a furnace containing silicon metal. When melted, the silicon wets the carbon and reacts with it to form SiC. The nascent SiC dissolves in the molten silicon and then precipitates out on the surface of the alpha SiC grains bonding them into a hard, strong body. The resultant body contains roughly 10% free silicon.

By careful choice of the types and amounts of added carbon, a body can be produced which contains up to 30% unreacted graphite as well as free silicon. This unreacted graphite gives limited self-lubricating properties to the reaction-bonded silicon carbide.

Sintered silicon carbide bodies containing no free silicon can be produced by a modification of the reaction-bonded process. Finely ground SiC powder is mixed with a sintering

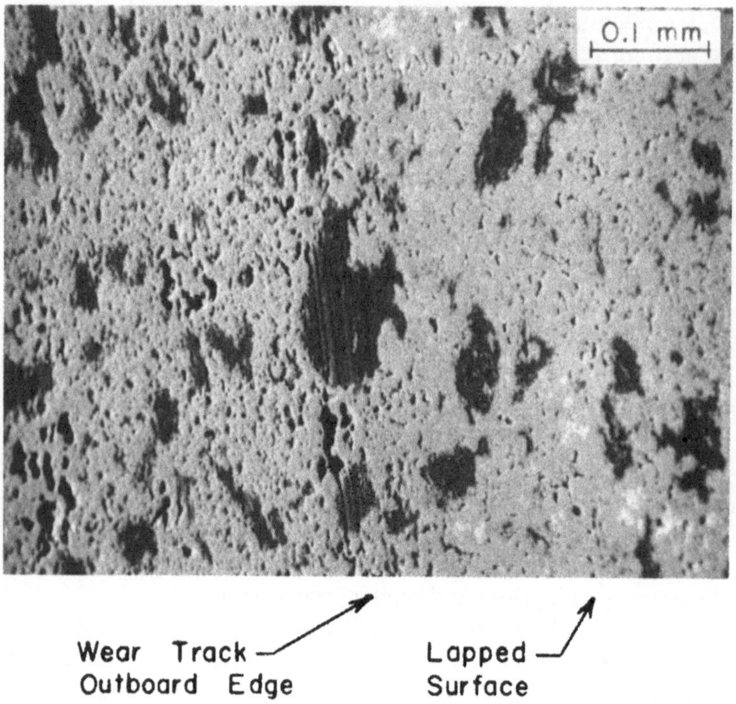

FIGURE 4.14. Scanning electron beam microphotograph of the polished rubbing surface on a siliconized carbon prepared by the CVR process.

aid such as yttria (Y_2O_3) and a fugitive temporary binder. Parts of the desired shape are pressed and then machined to slightly oversized dimensions. They are then sintered in an inert atmosphere at a temperature above 2000°C. The resulting solid silicon carbide body has outstanding corrosion resistance since it contains no free silicon.

Silicon carbide rings are rapidly supplanting sintered tungsten carbide rings as mating faces for carbon seal noses. They have better chemical resistance, are harder and in most cases cheaper than tungsten carbide rings.

XII. QUALITY CONTROL

The manufacture of carbon blanks suitable for the production of high-performance mechanical carbons is an exacting science drawing principally on the fields of chemistry and chemical engineering. Quality control should start with selection of reliable raw material vendors. Every shipment of raw material should be thoroughly inspected and given a lot number.

After each subsequent step in the process — mixing, molding, baking, and impregnation — the products should again be inspected for conformity to established standards. Doubtful parts should be discarded since it makes little sense to continue to add labor to parts of questionable quality.

Table 4.5 shows the build-up in value, expressed as dollars per pound, of a typical aircraft seal ring. This product is perhaps an extreme case due to the intricate machining required, but it illustrates clearly the need for a hard-nosed quality control group which unhesitatingly discards questionable material as soon as it is found.

TABLE 4.5

Value at Each Stage of Manufacture for a Typical Aircraft Seal Ring

Process stage	\$/lb of rings
Raw materials	0.50
After mixing	1.00
After molding	5.00
After baking	6.00
After impregnation	10.00
After rough machining	40.00
After finish machining	100.00
Inspected and ready for shipment	120.00

XIII. MANUFACTURING TIME CYCLES

The manufacture of high-quality carbon materials suitable for use in mechanical items is risky and time consuming. There are many process steps between purchase of the ingredients and shipping an intricately machined item. Yield for some of these steps is unpredictable, even under carefully controlled conditions. To be reasonably sure of shipping on the promised date, it is common practice to start with a substantial excess.

Other risk-reducing strategies include stocking sizeable quantities of semifinished blanks or even a supply of finished parts sufficient for several months. These latter steps tie up substantial amounts of capital. They also entail a risk to the manufacturer. If the buyer changes either his buying pattern or his part design, the manufacturer's inventory can become worthless. These risks are covered either by pricing or by long-term contractual arrangements.

REFERENCES

1. **Nightingale, R. E.,** *Nuclear Graphite,* Academic Press, New York, 1962, 22.
2. **Blackman, L. C. F.,** *Modern Aspects of Graphite Technology,* Academic Press, New York, 1970, 49.
3. **Johnson, N. J. and Nickerson, J. D.,** U.S. Patent 2,906,632, 1959.
4. **Bond, R. L.,** *Porous Carbon Solids,* Academic Press, New York, 1967, 155.
5. **Anon.,** Impregnation improves carbon-graphite, *Mater. Des. Eng.,* 60(3), 93, 1964.
6. **Paxton, R. R.** Carbon and graphite materials for seals, bearings and brushes, *Electrochem. Technol.,* 5, 174, 1967.
7. **Earp, F. K. and Hill, M. W.,**in Industrial Carbon and Graphite Papers Conf., London, 1957.
8. Union Carbide Corporation, British Patent 865,320, 1957.
9. **Paxton, R. R. and Beyer, G. I.,** U.S. Patent 3,342,627, 1967.
10. **Hough, R. L.,** Silicon Carbide, U.S. Patent 3,416,951, 1965.
11. **Sin'Kooskii, A. S.,** Analysis of reactions of silicon carbide formation from a gas phase on graphite, *Zasch. Pokrytiya Met.,* 1, 153, 1967.
12. **Gulden, T. J.,** Deposition and microstructure of vapor-deposited silicon carbide, *J. Am. Ceram. Soc.,* 51, 424, 1968.

13. **Lloyd, D. E., Howard, V. C. B., and Ceram, R. A.**, Fabrication and properties of large diameter tubes and coatings in pyrolytic silicon carbide, *Spec. Ceram. Proc. Symp. Br. Ceram. Res. Assoc. Iss.*, 4, 103, 1968.
4. **Cartwright, B. S. and Popper, P.**, Deposition of silicon carbide from methyltrichlorosilane, *Sci. Ceram.*, 5, 473, 1970.
15. Dow Corning Corporation, Surface Protection of Graphite Substrates, British Patent 1,118,056, 1968.
16. **Potter, H. N.**, Method of Producing Silicon Monoxide, U.S. Patent 875,675, 1907.
17. **Potter, H. N.**, Method of Making Silicon Carbide, U.S. Patent 875,693, 1907.
18. **Tone, F. J.**, Silicon Carbide, U.S. Patent 1,013,700, 1912.
19. **Tone, F. J.**, Silicon Carbide Article and Method of Making Same, U.S. Patent 1,058,800, 1913.
20. **Woerner, L. M., Kivela, W. C., and Quaal, G. C.**, Silicon Carbide Encased Articles and their Production, U.S. Patent 3,634,116, 1972.
21. **Nishikawa, T.**, Silicon Carbide Coated Shaped Bodies, German Patent 2,123,572, 1972.
22. **Nishikawa, T.**, Method of Producing Silicon Carbide Articles Having a Thin Layer Construction, U.S. Patent 3,726,737, 1973.
23. **Kozhevnikov, G. N., Vodop'yanov, A. G., Chufarov, G. I., and Sverdlovsk, G. I.**, Interaction of silicon monoxide with carbon, *Isv. Akad. Nauk SSSR Met.*, 4, 82, 1972.
24. **Knippenberg, W. F. and Verspui, G.**, Bearing Elements, U.S. Patent 3,591,243, 1971.

Chapter 5

MACHINING CARBON

I. INTRODUCTION

Manufactured carbon varies greatly in machinability. Low-strength graphized materials can be sawed and planed with woodworking tools,[1] but some mechanical carbons are so hard they must be ground with diamond wheels. The machining method used, moreover, may affect the initial friction and wear behavior of the resulting carbon part. Gupta and Cook[2] report that both surface finish, measured as rms, and the microtopography of the surface affect the interaction between contacting asperities of sliding surfaces. Campbell et al.[3] found that single-point tool machining of fine grain carbons disturbs the surface to a depth of 0.1 mm (0.004 in.). This subsurface disturbance extends to 0.3 mm (0.012 in.) on larger grained carbons and is caused by loosening the bonds of individual grains.

The purpose of this chapter is to discuss machining methods used to shape manufactured carbon and the tolerances generally achieved. Figure 1.1 shows a few typical machined parts.

II. SAWING

Most carbons can be sawed on band saws. A typical saw blade will have two teeth per inch (one tooth per centimeter) with a large throat between each tooth. Very little "set" is required because there is little tendency to bind the blade. Band saws are usually operated dry. Blade speed ranges from 0.5 m/sec (98 ft/min) for hard carbons to 5 m/sec (984 ft/min) for soft graphite. The stock feeds into the blade at speeds ranging from 0.1 mm/sec (1/4 in./min) for thick plates of hard carbon to over 4 mm/sec (9.4 in/min) for soft graphite. One must allow 0.15 mm (1/16 in) for the saw kerf. An accuracy of ± 0.8 mm (0.031 in) can be maintained on band saws.

Metal circular saws can cut to closer tolerances than band saws, but their use is limited to soft graphites. This limitation can be circumvented by the use of thin abrasive discs in lieu of metal ones. A typical disc such as Carborundum's C-368-R-BV should be run at 30 m/sec (5906 ft/min). Even so, slow cutting rates generally limit cutoff discs to specialized applications.

III. GRINDING FOR FLAT SURFACES

Wet grinding is commonly used to produce flat surfaces on plates and discs. Wheel grits are normally in the 32 to 60 mesh range (e.g., Carborundum's SC-46-P6-BA), and wheel speeds run to 25 m/sec (4921 ft/min). Stock feed rates of 50 mm/sec (9.8 ft/min) with removals up to 3 mm (1/8 in.) per pass can be achieved by the more powerful machines. On very soft graphites, high-speed planing mills may be preferable to wet grinders for producing plane surfaces. These mills run dry and thus avoid the possibility of contaminating porous graphite with water soluble impurities.

Thickness tolerances of ± 0.1 mm (0.004 in) are readily obtained with either wet grinding or dry milling. Parallelism of ground surfaces can be held to 400 μm/m length (0.0004 in./in.).

Flatness measurements are complicated by the fact that thin ground plates can be deflected by the forces used in making the measurement itself. On plates thicker than 6 mm (0.236 in.) total variation from a planar surface can be held below 1 mm/m (0.001 in./in.).

IV. EXTERNAL AND INTERNAL GRINDING OF ROUNDS

Where smooth surface finishes are required, grinding is the preferred method of shaping both hard and soft carbons. Dry grinding generally uses a coarse, more open grit wheel such as Carborundum's SC-46-P-6-BA, while wet grinding uses a wheel such as Norton's A6-M5VBE. In both cases, optimum wheel surface speeds range from 25 to 50 m/sec (4921 to 9843 ft/min). Surface removal rates range from 0.01 to 0.2 mm (0.0004 to 0.008 in.) per pass.

Tolerances of ± 0.05 mm (0.002 in.) are readily obtained. On diameters below 30 mm (1.18 in.) a modern, well-maintained machine can hold ± 2 μm (0.0001 in.) tolerance. Production rates on tolerances such as these are markedly lower; hence, costs rise sharply. Moreover, many carbon-graphites expand as much as 0.0005 mm/mm (0.0005 in./in.) when taken from a completely dry to a water-saturated state. Because of these two factors, the designer should be careful not to specify closer tolerances than actually needed for proper functioning of the device.

V. LATHE OPERATIONS (TURNING)

Many surfaces, such as internal O ring grooves, are best prepared using single-point tooling and a lathe. In this process, tool wear is the greatest problem since a worn tool affects both tolerance and finish. For any but the softest graphites, one should use diamond-, carbide-, or ceramic-tipped tools. A typical tip material is General Electric's Carboloy 833. Figure 5.1 shows a common way of dressing the cutting edge.

Rotational speed on the lathe should give a surface speed in the range of 2 to 5 m/sec (393 to 984 ft/min). Stock removal will generally range from 0.2 to 2.5 mm (0.001 to 0.01 in.) per pass. Lathe operations are generally carried out dry. On harder carbons, however, the use of water as a coolant can double the number of parts between tool sharpenings.

VI. DRILLING

Carbon can be drilled with high-speed twist drills or with spade drills. With twist drills, the greatest problem is clearing the carbon dust from the hole so it does not wedge the drill surface against the wall of the hole. The drill point should be dressed to a 90° angle and the drill frequently withdrawn for the hole to clear the chips. Alternately one can drill the part from the underside so the chips fall out.

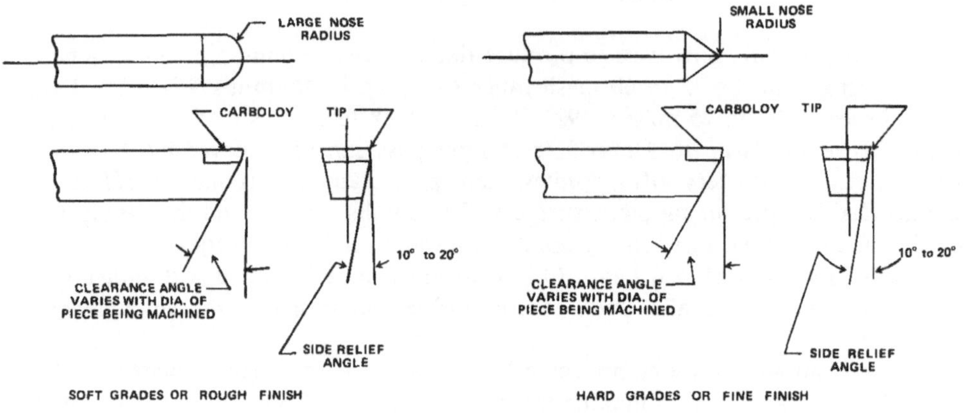

FIGURE 5.1. Lathe tools for machining carbon.

Drill speed ranges from 100 rpm for 25-mm (1-in.) holes in hard carbon to 20,000 rpm for a 0.5-mm (0.020-in.) hole in a soft graphite. Solid carbide drills are often preferred for holes below 6 mm (0.236 in.) despite their brittleness and higher cost. Where the hole depth to diameter ratio exceeds 10:1, gun drilling is often used.

Tolerances on drilled hole diameters can be held to ± 0.02 mm (0.001 in.) below 3 mm (1/8 in.). Above that, one should allow ± 0.05 mm (0.002 in.). Hole depth is difficult to control, and a tolerance of at least 0.2 mm (0.01 in.) should be allowed.

VII. REAMING

Holes in carbon can be reamed using conventional reaming tools. Except for the softer graphites, however, the reaming tool is quickly dulled.

VIII. TAPPING AND THREADING

Tapping or threading is not generally recommended for carbon parts. Tap life is short on all but the softest grades of graphite. Moreover, because of its low strain to failure, carbon is easily overstressed, thereby stripping the threads or breaking the carbon bolt. Where threading is the only practical fastening method, the coarsest threads possible should be used.

IX. BROACHING

Carbon can be broached using conventional metal broaching tools with carbide cutting edges. Because carbon is weak in tension, the parts must be well supported adjacent to the broach exit, otherwise there will be excessive chipping. Broaching is usually limited to use with soft graphites where tool wear is low.

X. HONING

Honing should be considered as a final operation where close tolerances and a good finish are required on an inside diameter. Using grits of 220 and 400 mesh in a medium-hard matrix and a well-filtered water coolant, one can obtain finishes as good as 0.15 μm rms (6 μin.) of fine-grain, resin-impregnated materials.

XI. LAPPING

Carbon surfaces can be lapped on machines such as the Norton Hypro-Lap to get flatness to 0.002-mm (0.0001-in.) total variation. Depending on the speed of cutting desired and the finish sought, lapping stones vary from 120 grit to 320 grit. The latter produces a scratch free finish known to the trade as "commercial lap".

Lapping operations can be wet or dry. Clean water free of suspended or dissolved solids is used for wet lapping. In general, the use of coolants or lapping compounds should be avoided since these may be imbibed by porous carbons to the detriment of their running properties. Fully impregnated impervious seal grades of carbon, of course, are not bound by this restriction.

Flatter surfaces with a higher polish may be obtained with the Norton Seal-O-Lap, the Crane Lapmaster, or hand lapping on a cast iron flat carrying a 600-mesh paper such as Carborundum's Aloxite A600 LB 981F. Flatness to two helium light bands per inch (20 μ in./in. or 20 μ m/m) with a mirror finish is obtainable on premium seal grade mechanical carbons. Figure 5.2 summarizes machining methods for carbon.

Operation	SOFT CARBON Surface Speed in Ft./Min.	SOFT CARBON Type of Wheel or Tool Used	MEDIUM HARD CARBON Surface Speed in Ft./Min.	MEDIUM HARD CARBON Type of Wheel or Tool Used	HARD CARBON Surface Speed in Ft./Min.	HARD CARBON Type of Wheel or Tool Used
Sawing Band	500	Standard Steel Blade	350	Standard Steel Blade	180	Standard Steel Blade
Cut-off Wheels	6,500	"N" 37C46-O4E "C" C-368-R-BV	6,500	"N" 37C46-04E "C" C-368-R-BV	6,500	"N" 37C46-04E "C" C-368-R-BV
(Wet)	4,500	"C" C-368-R-BV	4,500	"C" SC46-P6-BA	4,500	"C" SC46-P6-BA
Grinding (Thkns.) Dry	7,000	"	7,000	"	7,000	"
Grinding (External Wet)	6,500	"C" C-60-34-VE "N" A-60-M5VBE	6,500	"C" C-60-34-VE "N" A-60-M5VBE	6,500	"C" C-60-34-VE "N" A-60-M5VBE
Grinding (Internal Wet)	6,500	"C" C46L4VGC	6,500	"C" C46L4VGC	6,500	"C" C46L4VGC
Turning (Dry)	400 to 1,000	Carboloy® 883 tip	400 to 1,000	Carboloy® 883 tip	400 to 1,000	Carboloy® 883 tip
Drilling (Dry)	600 RPM to 2400 RPM	High Speed Twist Drill Point Angle 90°	150 RPM to 1800 RPM	Carboloy® tipped masonry drill point angle 120°	130 RPM to 1800 RPM	Carboloy® tipped masonry drill point angle 120°
Honing	No advantage on these grades		300 RPM	Grit 220 or 280 Bond-medium hard	300 RPM	Grit 220 or 280 Bond-medium hard
Lapping Parallel		"N" 120J9V or "N" 180 J9V		"N" 120J9V or "N" 180J9V		"N" 120J9V or "N" 180J9V
Lapping Comericial		"N" 320K9V		"N" 320K9V		"N" 320K9V
Lapping Light Band		"N" 500J9V		"N" 500J9V		"N" 500J9V
Lapping Seal-O-Lap	Spindle 240 RPM Wheel 15 RPM	"N" 320J or K "N" 500J or K	Spindle 240 RPM Wheel 15 RPM	"N" 320J or K "N" 500J or K	Spindle 240 RPM Wheel 15 RPM	"N" 320 J or K "N" 500J or K
Lapping Hand		A600LB981F Aloxite® Resinoid		A600LB981F Aloxite® Resinoid		A600LB981F Aloxite® Resinoid

NOTE: "N" - Norton Wheel "C" - Carborundum

FIGURE 5.2. Summary of carbon machining methods.

XII. METALWORKING OPERATION

Since carbon does not melt, soften, or flow, many familiar metalworking operations cannot be done. Carbon cannot be forged, stamped, or bent. It cannot be welded, but it can be soldered provided it is first plated with silver or copper. The soldered joint has little strength, however, because the bond between the carbon and the plating is a purely physical one rather than a metallurgical one. If there is no need for electrical conductivity across the bond, carbon may be cemented with any one of a number of adhesives.

XIII. CEMENTED CARBON

Carbon can be cemented to itself or to metal parts. If the cemented assembly must retain its integrity over a range of temperatures, care must be taken to match the thermal expansion coefficient of the cemented partners. Many carbons, it will be recalled, have different thermal expansion with and across the grain; hence, care must be taken even with assemblies of a single grade of carbon if one is to obtain a good match.

Most organic cements are formulated from a resin thickened with finely divided carbon. A catalyst or curing agent for the resin is often supplied in a second container. Resin systems in common use are phenolics, furanes, and epoxies. Because of their high coking yields, phenolics and furanes are preferred if the joint is to be heated to carbonizing temperatures.

For maximum strength, cemented joints should be thin, 0.1 to 0.3 mm (0.004 to 0.012 in.), and free from voids. The contacting surfaces must be smooth, parallel, free of dust and contaminants, and of low surface porosity. Highly porous graphites tend to imbibe the resin and leave a weak joint. Since carbon holds chemisorbed water tenaciously, it is good practice to bake out the specimens and cool them in a desiccator prior to applying cement. This will minimize gas evolution and bubble formation in the joint.

Curing the cemented joint usually requires baking to 150°C (302°F). The joint should be clamped lightly, 35 kN/m^2 (5 $lb/in.^2$) and the assembly heated slowly, 10°C (18°F) per hour. The cured joint can be carbonized or even graphitized by heating to elevated temperature in a reducing atmosphere.

Strength of carbon-carbon joints can readily reach 75% that of the substrate carbon. Carbon-metal joints generally fail at the cement-metal interface, but their strength is 50 to 75% that of the carbon itself. Carbon-carbon joints which have been carbonized are nearly as strong as uncarbonized joints. Electrical resistivity of carbonized joints ranges from 40 to 200×10^{-4} Ω cm.

XIV. DUST COLLECTION

A discussion of carbon machining would not be complete without a few words about dust collection. So far as is known, graphite (carbon) dust is not toxic, and it does not form explosive mixtures in air (as does coal dust, for example). However, it is a nuisance, and when mixed with a little oil it will quickly wear away bearings and lathe ways. Thus, a good dust collection system is a must for any shop proposing to machine carbon. An air velocity of at least 2.5 m/sec (492 ft/min) should sweep across the cutting area of each dry machining device to entrain the dust as it is formed. This dust laden air is then circulated through bag filters and returned to the shop. Velocity in the ductwork should be near 25 m/sec (4921 ft/min) to avoid settling and clogging of the lines.

Many carbons used for mechanical applications contain additives such as copper, babbitt, antimony, or inorganic fluoride and phosphate salts. Dust containing these materials should be treated with the respect due any aerosol of potentially toxic materials.

Dust-laden water from wet grinding operations should be routed to a holding pond big enough to permit 8 hr of dwell time. This generally will allow more than 90% of the solids to settle as a sludge. The clarified overflow from this settling basin can then be discharged to the plant system for handling liquid wastes without fear of line plugging from carbon sludge.

For direct discharge into surface streams it is desirable to get the level of suspended solids below 50 mg/ℓ. To reach these levels, flocculating agents are generally added to speed sedimentation of fine carbon particles.

REFERENCES

1. **Nightingale, R. E.,** *Nuclear Graphite,* Academic Press, New York, 1962, 54.
2. **Gupta, P. K. and Cook, N. H.,** Statistical analysis of mechanical interaction of rough surfaces, *J. Lubr. Technol. ASME Trans.,* 19, January 1972.
3. **Campbell, J. G., Dennington, M. G., and Shirodkar, D. N.,** The use of carbon and graphite in slides/counterfaces systems in the field of mechanical engineering, in *Proc. Industrial Lubr. Tribology Symp.,* London, November, 1969, Scientific Publications (G.B.) Ltd., 1970, 121.

Chapter 6

CARBON IN MECHANICAL SEALS

I. INTRODUCTION

Rubbing elements in face and circumferential seals represent the most important single use for carbon in mechanical devices. Reviews of the literature on operation and design of seals have been prepared by King,[1] Stair,[2] Ewbank,[3] Elwell et al.,[4] and later by Koenig.[5] Altogether these reviews survey over 2500 papers and unclassified reports issued through 1967. Mayer[6] wrote a book on mechanical face seals which was published in 1963; an English language edition of his book was released in 1969.

The information available in the literature provides the seal designer an extensive background on the theory underlying mechanical seals. There are also abundant details on the design, testing, and application of mechanical seals. Relatively little, however, has been published on the selection of rubbing materials to be used in mechanical seals. The purpose of this chapter will be to help the seal designer specify the primary rubbing elements for his seal assembly.

Figures 6.1 and 6.2 show cross sections of typical face seal assemblies together with the standard nomenclature for their parts. Further details on seal nomenclature are given by Schoenherr[7] and Brown.[8] Dahlheimer's book[9] contains excellent illustrations of common seal types, while Cawley and Duncan[10] and Giles[11] describe more complex assemblies. Figure 6.3 shows a cross section of a typical circumferential seal assembly. A photograph of the components of a typical face seal is shown in Figure 6.4.

II. INTERACTION OF CARBON AND MATING FACE

In any mechanical seal, the critical area is the rubbing interface between the carbon seal nose and its mating metal or ceramic wear plate. These two rubbing members must interact to maintain a smooth, low-leakage junction with a low coefficient of friction and a low wear rate at the sealing face.

Nearly all seal designs call for the carbon nose to absorb the wear while the harder mating material presents a flat smooth surface on which the carbon sealing face rides. The first principle of good design in these cases is to have the harder wear plate extend beyond the sealing face of the carbon seal nose on both inner and outer diameters. No matter how great the misalignment of shaft runout, the carbon nose must never overhang a hard wear plate. Failure to observe this precaution invariably results in the hard wear plate cutting a recess in the softer carbon seal nose.

Within the first few hours of seal operation, a transfer film should form on the mating metal or ceramic wear plate. When the film forms, the coefficient of friction and the wear rate drop. Wilkes[12] describes the role of this film in reducing wear rates. Hershey[13] recognized its effect in his description of a seal test method which discards the wear observed during the first day and uses that of the following four days to project seal life. Figure 6.5 shows the drop in friction coefficient observed by Paxton and Shobert[14] during the first 100 hr of a test in which carbon is run against a cast-iron wear plate.

The operational life of a seal depends on this transfer film. If it fails to form — or if it is abraded away as rapidly as it forms — the wear rate will be high and seal life short. On the other hand, if there is too much filming, a black, high-friction wear track forms, and considerable heat is generated. Sometimes the film thickens to the point where it loses adhesion and flakes off the wear plate. When this occurs, the seal

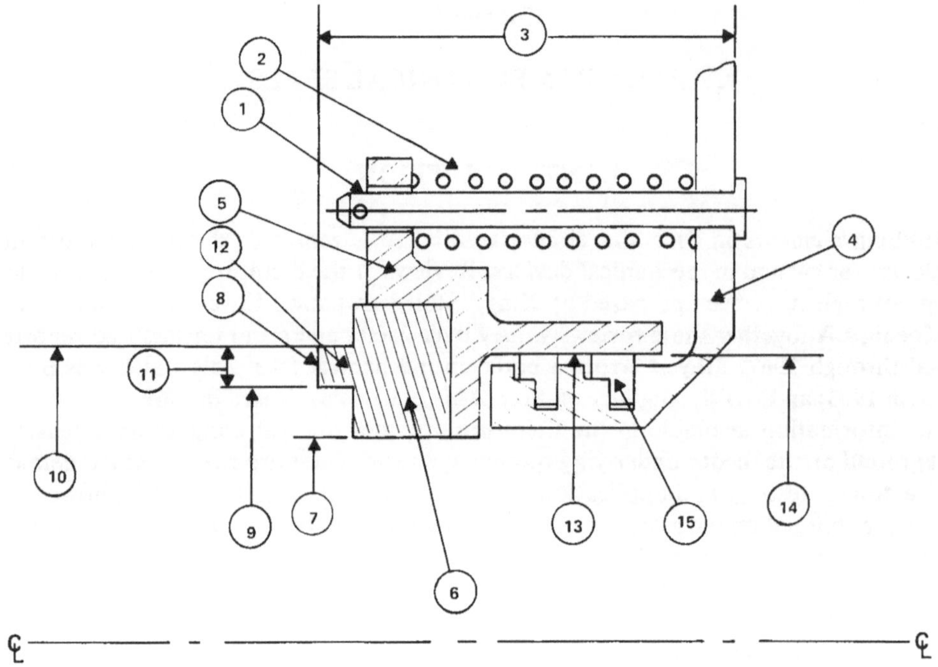

Legend:

1 – Torque Pin
2 – Coil Spring
3 – Free Length
4 – Piston-Ring Carrier
5 – Primary-Seal Carrier
6 – Primary-Seal Ring, Carbon
7 – Seal Bore
8 – Sealing Face
9 – Sealing-Face Inner Diameter
10 – Sealing-Face Outer Diameter
11 – Sealing-Face Width
12 – Seal Nose
13 – Secondary-Seal Land
14 – Secondary-Seal-Land Diameter
15 – Piston Ring

FIGURE 6.1. Mechanical face seal assembly with pin-type rotation stop. (From *ASLE Special Publication SP-1*, American Society of Lubrication Engineers, Park Ridge, Ill., 1969. With permission.)

will leak until the debris clears the gap; then the seal closes and runs normally for hours or days before the cycle is repeated.

At high rubbing speeds and nose pressures, this film-formation, film-loss cycle can have a period of minutes or even seconds. Abar[15] describes carbon nose temperature fluctuations of 14°C (25°F) within a 3.5- to 4.8-sec period. These temperatures were measured 0.8 mm (0.030 in.) back from the rubbing interface; hence, they greatly understate temperature variations at the rubbing face. Heat checking of the metallic wear plate and high carbon nose wear rates were also observed.

Factors that control the rate of transfer film formation are poorly understood. Savage[16] and Smith et al.[17] describe the role of water vapor in this process. Snow et al.[18] and Buckley[19] show that oxygen and metal surface oxides are also important. Paxton and Shobert[20] discussed the different appearance of this transfer film after tests which showed high and low carbon wear rates.

In an operating seal there must be a dynamic balance between the rate of formation of transfer film and the rate at which it is abraded away. Since the rubbing speed, face pressure, temperature, and fluid environment were largely determined by the designer and the application requirements, it became the task of the carbon producer to supply a carbon that forms a proper transfer film. Over the years, each manufacturer of seal carbons has developed his group of 10 to 50 grades to satisfy the range of operating conditions usually encountered.

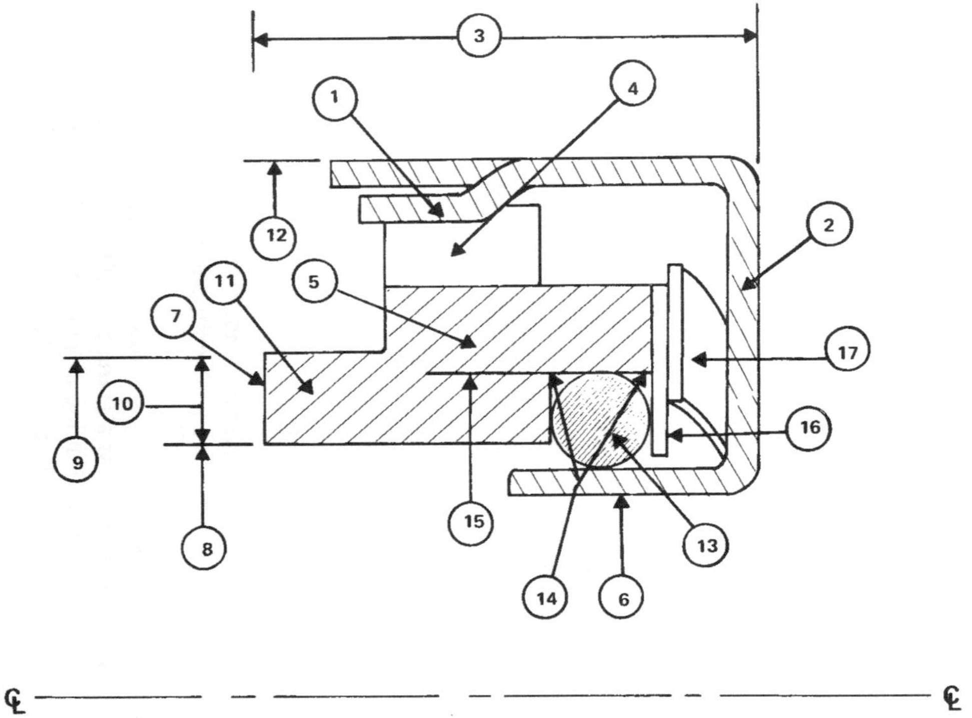

Legend:

1. Anti-Rotation Device
2. Cup
3. Extended Length
4. Primary-Seal Carrier
5. Primary-Seal Ring, Carbon
6. Seal Bore
7. Sealing Face
8. Sealing-Face Inner Diameter
9. Sealing-Face Outer Diameter
10. Sealing-Face Width
11. Seal Nose
12. Seal Outer Diameter
13. Secondary Seal (O-Ring)
14. Secondary-Seal Land
15. Secondary-Seal-Land Diameter
16. Washer
17. Wave Spring

FIGURE 6.2. Mechanical face seal assembly with shear-nitch rotation stop. (From *ASLE Special Publication SP-1*, American Society of Lubrication Engineers, Park Ridge, Ill., 1969. With permission.)

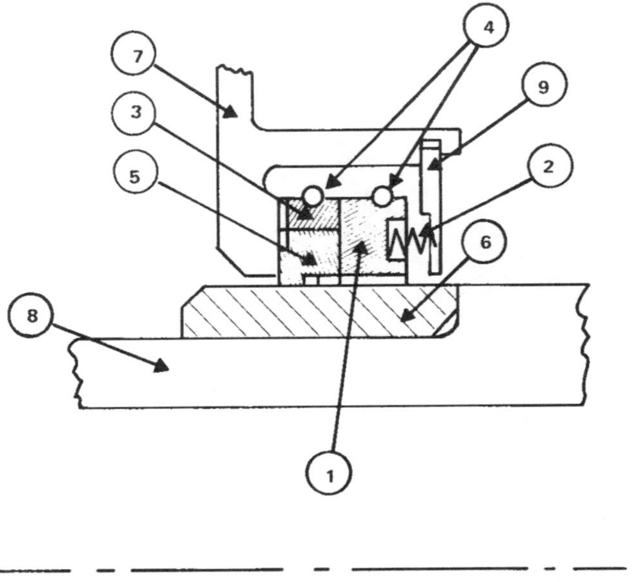

Legend:

1. Back Ring, Carbon
2. Coil Spring
3. Cover Ring, Carbon
4. Garter Springs
5. Primary-Seal Ring, Carbon
6. Runner
7. Seal Housing
8. Shaft
9. Spring Adapter

FIGURE 6.3. Circumferential seal assembly. (From *ASLE Special Publication SP-1*, American Society of Lubrication Engineers, Park Ridge, Ill., 1969. With permission.)

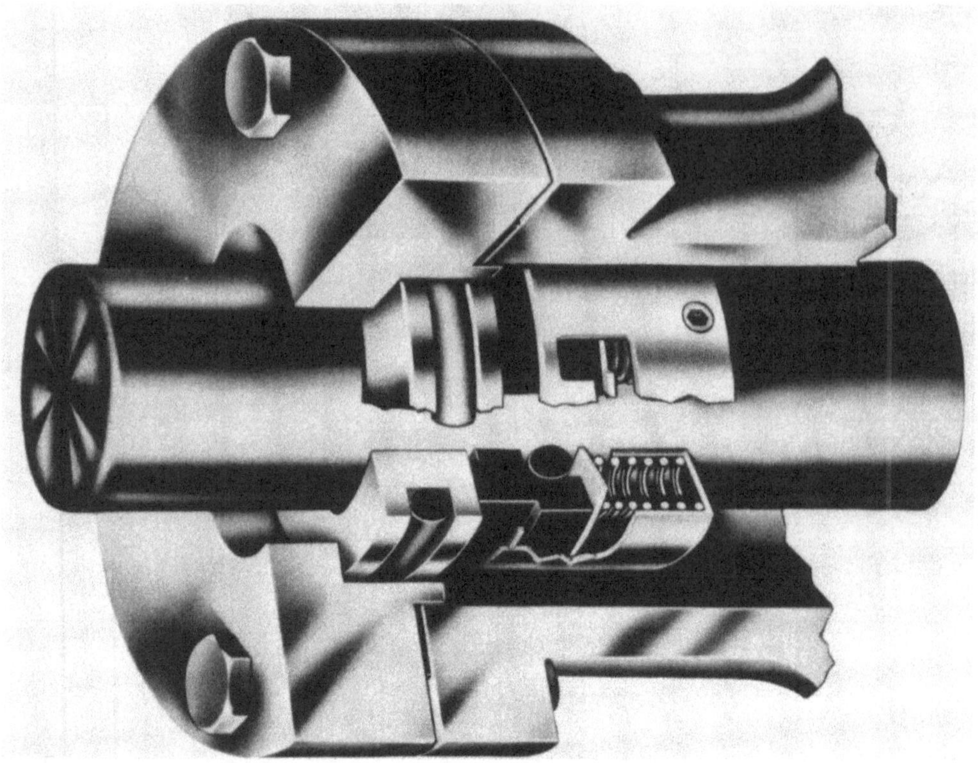

FIGURE 6.4. Mechanical face seal assembly. (Courtesy of John Crane Company.)

III. CHOOSING THE CORRECT CARBON GRADE

Listed below are the major factors which must be considered in selecting the primary rubbing elements:

1. Mating face material, hardness, and surface finish
2. Fluid present at the seal interface (higher pressure fluid)
3. Ambient fluid (lower pressure fluid)
4. Operating temperature and peak temperature
5. Rubbing speed
6. Pressure gradient across seal face
7. Maximum (and average) face pressure
8. Permissible leak rate
9. Life expected
10. Size and configuration of carbon parts

IV. WEAR PLATE MATERIALS

Relatively few wear plate materials are used commercially. Choice among them is usually dictated by economic considerations, by corrosion problems, and by the hardness required, factors discussed in detail by Mayer.[21]

Fine-grain cast iron (Meehanite®) is used in high-volume, low-cost seals such as those on water pumps and gear trains.[22] Because it has a textured surface, cast iron usually demands a seal carbon with relatively high filming action. The seal carbon should also be low in abrasiveness or "polishing action", as it is often termed.

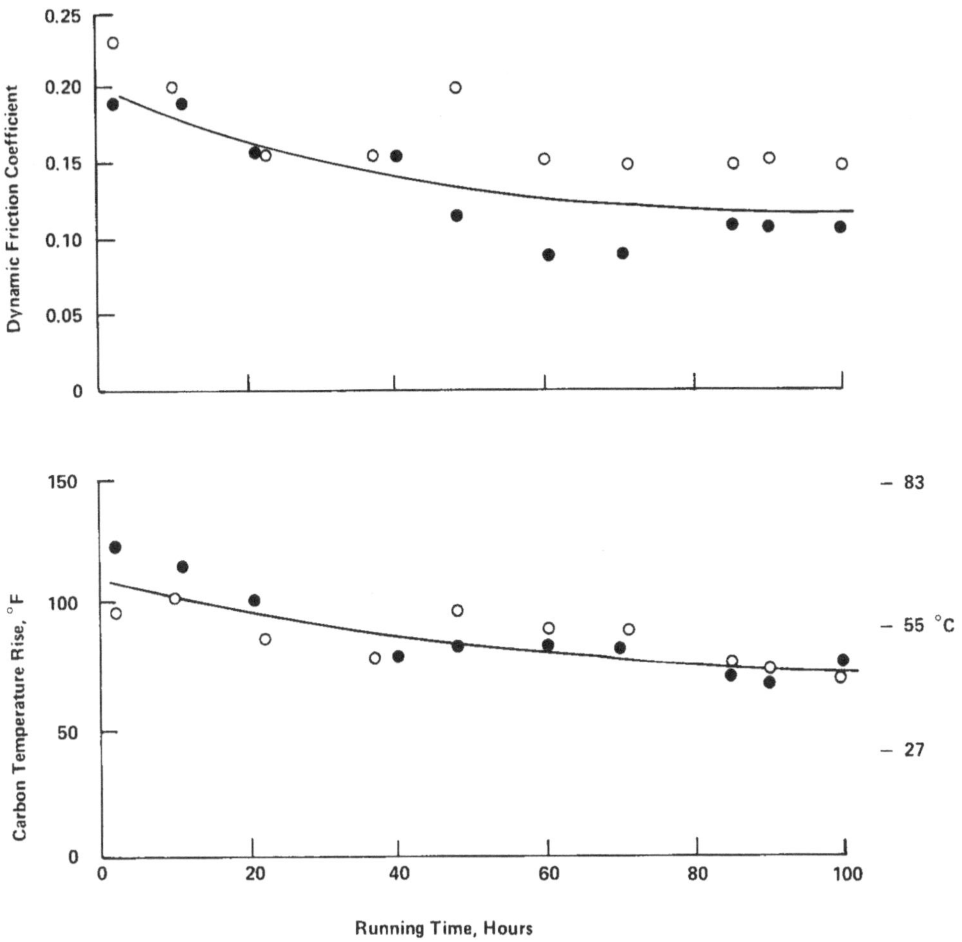

Operating Conditions:

Rubbing Speed	- 2.29 m/sec (450 ft/min)
Running Dry-Ambient Air	- 16-32 °C (60-90°F)
Face Pressure	- 0.14 MN/m^2 (20 psi)
Mating Face	- Cast Iron
○ ● Duplicate Runs	

FIGURE 6.5. Friction coefficient and temperature rise (above ambient) decline as run progresses.

Cast iron tends to rust in many applications and hard, chrome-plated steel is often substituted. Chrome plate, however, has a dense adherent oxide film which quickly builds a thick carbon transfer film. Chrome plate is also harder than cast iron and less subject to abrasive wear. Because of these properties of chrome plate, carbons with less film-forming tendency and more polishing action generally give lower wear rates and lower friction. (Increased polishing action in a manufactured carbon is readily obtained by substituting coke flour for a portion of the graphite filler.)

Table 6.1 illustrates this effect with data obtained on the "drill press" test stands of Paxton and Shobert.[14] These units test the carbon-metal couple as a thrust bearing. Under the conditions of this test, carbon grade P-692 shows the lower wear rate against cast iron, whereas the more abrasive carbon, P-658-RC, gives lower wear against chrome plate.

Hardened tool steel, hardenable stainless steels such as 17-4 PH, and the high nickel

TABLE 6.1

Effect of Mating Face Material on Friction and Wear Rate of Two Seal Grade Carbons

Carbon grade[a]	P-692	P-658-RC
Graphite content (%)	67	30
Hardness (Scleroscope)	75	90
Average wear rate mm/100 hr (in./100 hr)		
On cast iron	0.37 (0.015)	0.40 (0.016)
On chrome plate	0.52 (0.021)	0.46 (0.018)
Coefficient of friction		
On cast iron	0.12	0.18
On chrome plate	0.15	0.13
Number of tests averaged		
On cast iron	6	10
On chrome plate	11	20

Note: Test conditions:

nose load,	0.14 MN/m² (20 lb/in.²);
rubbing speed,	2.7 m/sec (525 ft/min);
test duration,	100 hr;
ambient temperature;	24°C (75°F);
ambient fluid,	room air both sides;
PV;	0.38 MPa·m/s (10,500 lb ft/in.² min);
Surface finish (metal),	8 to 12 μ. in. rms;
flatness (both members),	< 3 helium light bands per inch.

The carbons are P-692 (similar to grade P-29, Table 4.2) and P-658-RC (similar to grade P-5, Table 4.1, with added polishing action and thermoset resin impregnation).

[a] Material designations are those of Pure Carbon Company, Inc., St. Marys, Pa.

alloys such as the "Stellites" and "Hastelloys" tend to behave like chrome plate. All these materials form hard, adherent oxides that hold the carbon transfer film very well; hence, they are best mated with the harder carbons that have more polishing action.

High-density, fine-textured, fused (or bonded) alumina makes an excellent carbon mating face where water or corrosive chemicals must be sealed. Even when alumina is polished to a finish of 8 to 16 μin., it has a definite "tooth"; therefore, one would expect it to behave like cast iron and prefer a relatively soft carbon with good film-forming action. Two other factors are at work, however. The relatively large hard aluminum oxide grains tend to abrade the softer carbons. More important is the tenacity of the alumina-carbon bond holding the transfer film. In this connection, it is noteworthy that the aluminum-oxygen chemical bond is twice as energetic as the iron-oxygen bond.

Figure 6.6 illustrates the preference of alumina for the harder carbon. Each data point represents a 100-hr test sealing demineralized water at system pressures ranging from 0.14 MN/m² (20 psig) to 0.62 MN/m² (90 psig), and a temperature of 50°C (122°F). A commercially available 92-mm (3 5/8-in.) bore seal assembly equipped with a Coors AD-85 wear plates was used. The wear plates were relapped between tests to a flatness of 3 helium light bands per inch and to a surface finish of 0.15 to 0.75 μm rms (6 to 30 μin.). New carbon noses were used for each test.

I suspect the scatter in the data at low PV levels is caused by intermittant formation of a full hydrodynamic film. When a water film separates the faces, the wear rate, of course, drops to very low values. As face loading increases, the hydrodynamic film becomes thinner, and the inherent wear characteristics of the rubbing pair starts to dominate.

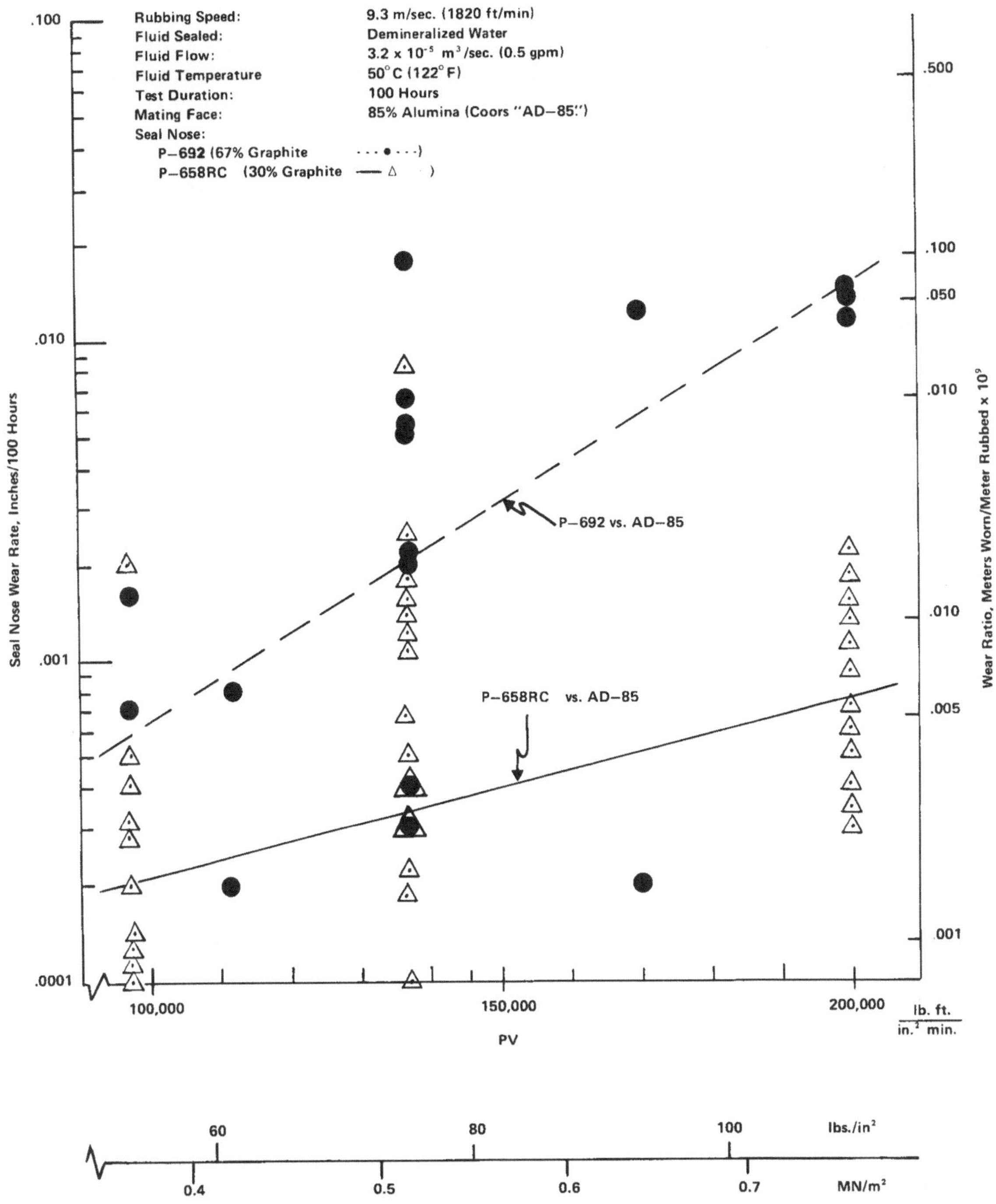

Closing Force per Unit of Seal Face Contact Area

FIGURE 6.6. Wear rate of two seal nose materials.

One can draw several important lessons from Figure 6.6. Selection of the correct rubbing pair becomes important only when it is difficult to generate and maintain a hydrodynamic film at the rubbing interface. A wide variety of carbon nose and wear plate materials can be used with equal success if one is sealing clean, high-viscosity liquids, and if the closing force at the seal face is low and the rubbing speed high. If, on the other hand, we are sealing low-viscosity liquids or liquids containing suspended or dissolved solids, film formation is more difficult. High closing loads on the seal face, poor surface finishes, low rubbing speeds, or frequent stops and starts compound the problem. Under these conditions, selection of materials for the primary rubbing pair becomes a critical factor in determining seal life.

In general, carbon wear rates go down as mating face hardness goes up. Piehn[23] reports the wear rates of babbitt- and bronze-impregnated carbon seal noses were reduced to one tenth their former values when he changed the wear plate from a nickel-iron alloy to cemented tungsten carbide. Wear of the metallic mating face, too, was reduced by the same fraction. Voitik[24] also found a one to one correlation between carbon wear rates and mating face wear rates. Mating face hardness values below 40 on the Rockwell C scale generally go hand in hand with short seal life.

A realization of this importance of mating face hardness led to the development of siliconized carbon mating faces. Figure 6.7 shows that the wear rate of P-658-RC seal noses was reduced by an order of magnitude when siliconized carbon was substituted for alumina. The data for this figure was obtained on the same apparatus[25] as that for Figure 6.6.

Even lower wear rates can be expected by ultrahard carbides such as solid silicon carbide and solid boron carbide.

V. HYDRODYNAMIC LUBRICATION

As we showed earlier, seal nose wear rates are greatly reduced by the presence of an interfacial hydrodynamic film.[26] This film is generally less than 1 μm (40×10^{-6} in.) thick. It minimizes solid-solid contact, reduces the friction coefficient, and removes heat generated between the contacting faces. It would be desirable, therefore, to pick a grade which enhances the development of this film.

Kojabashian and Richardson,[27] Hamilton et al.,[28] and Anno et al.[29] have investigated the role of microirregularities on the carbon and mating face in generating the necessary pressures to maintain a hydrodynamic film. In theory at least, the grain size (texture) of the seal nose carbon should be selected to complement the viscosity and density of the contact fluid. Moreover, asperity height should not exceed film thickness. This theory has resulted in a definite trend toward finer textures among commercial grades of seal carbons. The restraining factors of the trend are the high cost of preparing very fine carbon and graphite fillers and baking problems caused by the higher binder levels required to coat the finer filler particles.

Flatness of the seal and mating face also affects the hydrodynamic film as do axial shaft vibrations, eccentric running, and angular misalignment of the faces. Sneck,[30,31] Findlay,[32] and Lohou and Godet[33] investigated the role of these factors in generating cavitation of the hydrodynamic film. Pape[34] built an ingenious seal tester designed to minimize these factors. Three capacitive proximity probes equally spaced around the circumference were built into the stationary mating face. Both metal sealing surfaces were flat to within one light band (0.3 μm or 11 μin.). A detailed interferometric survey showed waviness in the circumferential direction roughly characterized by a two-cycle sine wave with a peak to valley amplitude of 0.1 μm or 4 μin. When the faces were mated and operated as a seal, this waviness was amplified by a factor of three. Film thickness, as measured by the proximity probes, cycled twice each revolution ± 0.3 μm (11 μin.). This cycle was superimposed on an average thickness which ranged from 0.4 to 0.7 μm (15 to 30 μin.) depending on the rotating speed, closing force, imposed pressure gradient across the face, and viscosity of the fluid sealed. By commercial standards, Pape's seal faces would be considered very flat, yet there was enough wav-

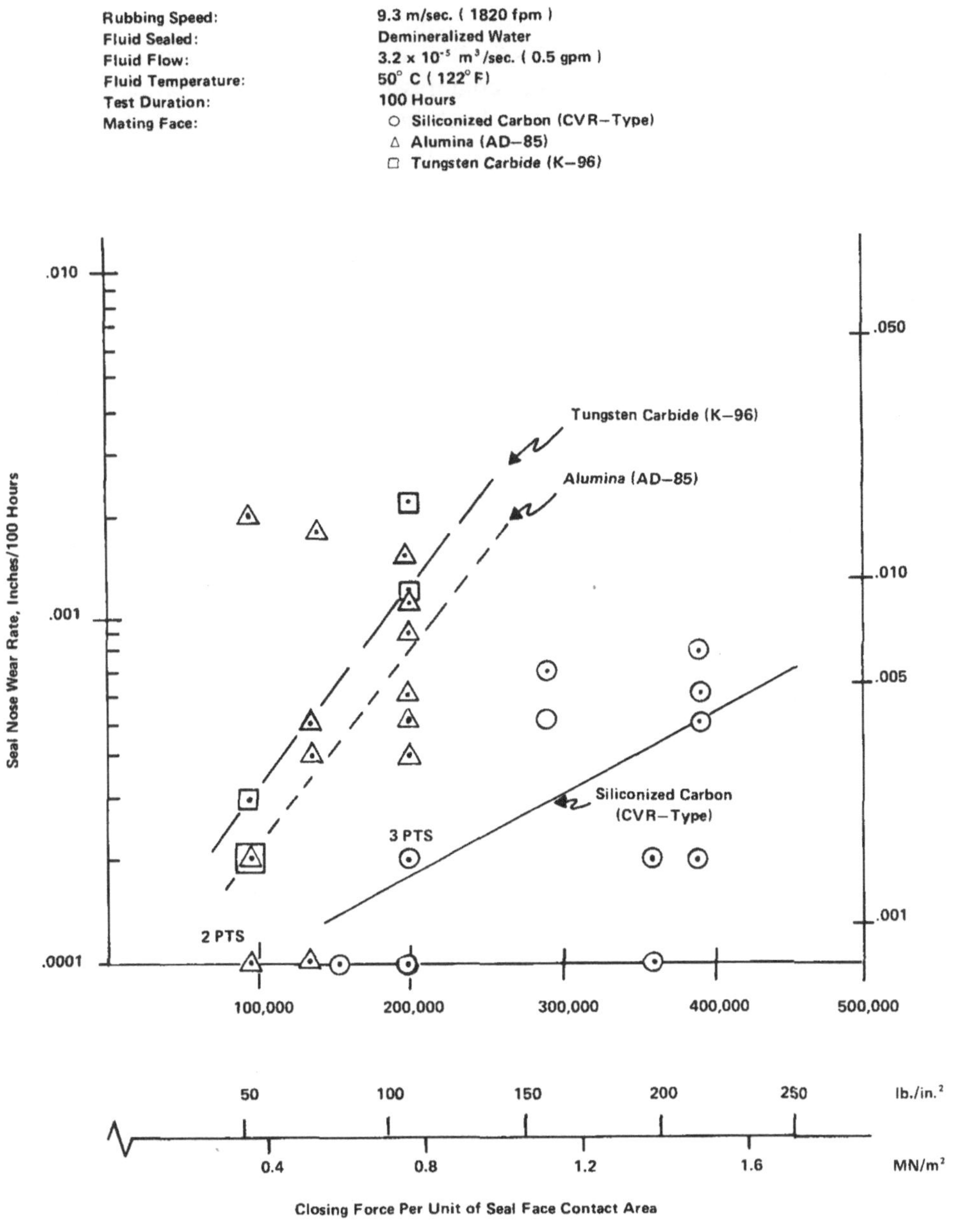

FIGURE 6.7. Wear rate of P-658-RC seal noses vs. various mating face materials.

iness in them to be significant relative to the total hydrodynamic film thickness. Carbon is a compliant, low-modulus material compared to Pape's metals; hence, it seems likely that the augmentation of waviness he observed would be greater in the typical seal with its carbon nose.

Pape also shows that the pressure of a fluid wedge converging radially in the direction of leakage promotes hydrodynamic film formation. For example, a wedge 0.8 μm (30 μin.) across a 5-mm (0.2-in.) face coupled with a two-cycle circumferential waviness of 0.12 μm (5 μin.) suffices to account for the thickness of his hydrodynamic film over the full range of operating conditions explored. Similar findings have been reported by Stanghan-Batch and Iny.[35]

In viewing these dimensions, it should be kept in mind that manufactured carbon is a low-modulus material, a typical modulus being 20 GN/m^2 (2.9×10^6 lbs/in.^{2}). A pressure gradient of 2.8 MN/m^2 (400 lbs/in.^{2}) across a seal face is enough to produce a 1.4 μm (55 μin.) wedge by differential compression of a carbon seal nose only 10 mm (1/4 in.) high. I suspect much of the "art" in the design of commercial seals lies in the clever use of this conformability of carbon to promote the formation of a hydrodynamic film.

VI. CHEMICAL COMPATIBILITY

The chemical properties of the fluid being sealed also play a major role in carbon grade selection. Carbon itself is one of the most inert materials available to the engineer. It is completely unaffected by acids, caustics, or solvents. Only strongly oxidizing agents such as 65% HNO_3 or sodium hypochlorite will attack the carbon, and some of the newer seal grades are 100-fold more resistant than those available in the mid-1960s.

Most seal carbons, however, contain impregnants to render them impervious. If the sealed fluid reacts with this impregnant, the seal nose may start to leak. The film-forming properties of the seal nose material are likely to be drastically altered also. Major manufacturers of seal carbons supply chemical compatibility charts to aid the user in his selection of carbon grades. They also maintain engineering staffs to aid the designer in grade selection for unusual applications.

Appendix C shows chemical compatibility for five seal grades manufactured by the Pure Carbon Company of St. Marys, Pa.

VII. SUSPENDED AND DISSOLVED SOLIDS

Liquids containing suspended or dissolved solids pose special problems to face seals. Suspended particles work their way into the interface between the carbon and its mating face. Leakage occurs when the faces are wedged apart. Eventually, the intruding particle is either ground up or embedded in the softer of the two contacting faces. In the latter case, it frequently cuts a groove in the contacting face before the particle is destroyed.

One approach to coping with suspended solids is to use the hardest materials available for both seal nose and mating face. Greiner and Bussick[36] describe the use of cemented tungsten carbide running against itself for sealing high-temperature, heat-transfer fluids where suspended pipe scale and carbon particles are present. The hard faces, coupled with strong seal closing forces, grind up the intruding particle before significant leakage can occur. Alternately, one may choose a seal-nose material which is soft and resilient enough to embed the foreign particle. A hard mating face gradually wears the particle away without opening the seal gap and causing leakage.

A recent development has been the use of a siliconized carbon seal nose running against a siliconized carbon mating face. Mechanical seals so equipped have been reported to have a fivefold to tenfold increase in life where abrasion is a problem. Siliconized carbon has the advantage over tungsten carbide in that it can run dry without heat checking. It is also more resistant to chemical corrosion. Most important, however, is the low modulus of siliconized carbon. This permits the nose and mating face to deform as needed to maintain perfectly parallel rubbing surfaces. Both wear rate and power loss are reduced with this combination of rubbing faces.

Dissolved solids present a different type of problem. Evaporation of the solvent at the rubbing interface leaves them behind. If allowed to build up, the deposited solids lead to high friction, high wear, and leaking. Janetz[22] and Lazau et al.[37] report that the presence of 1200 to 10,000 ppm of dissolved sodium chromate in circulating hot water increased the wear rate of the carbon seal nose tenfold. Trytek[38] describes special seal designs which permit the sealing of 121°C (250°F) water containing 2000 ppm of sodium chromate. In general, seals for fluids containing dissolved solids require an extremely hard mating face and a carbon (or carbide) seal nose with high polishing action.

Pfautz et al.,[30] Shevchenko and Brown,[40] and many other researchers have described the build-up of tar and coke at the interface of seals that are used to contain hot oil. A build-up of deposits at either the rubbing face or axial sliding clearance can cause gross leakage; hence, it must be avoided to all costs. The general method for preventing these build-ups is to use a high closing pressure and a hard carbon nose with a low coefficient of friction and enough polishing action to keep the interface clean. One may also purge the seal cavity with clear fluid as described by Tankus[41] and others.

VIII. OPERATING TEMPERATURES

The temperature of a carbon body has a profound effect on its filming action and, hence, on its coefficient of friction and its wear rate. In general, increased temperature results in increased wear rates. Johnson et al.[42] found the wear rate of carbon riders increased slowly up to the 204 to 316°C (400 to 600°F) range, after which it rose rapidly. Campbell et al.[43] points out that oxidation of the carbonized bond between filler grains becomes significant above 260°C (500°F) and that this, no doubt, contributes to higher wear rates. Brown et al.,[44] when testing carbon for use in aircraft jet engine seals, also reported increasing wear with increasing sealed gas temperatures up to 538°C (1000°F). Brown also observed a tenfold difference among carbon grades insofar as actual wear rates were concerned. In this connection, Paxton and Shobert[14] have reported that impregnation of graphite with an unspecified metal salt reduced its wear rate at 316°C (600°F) to 1/25 the former level. From these and other reports,[45] it is clear that the wear rate of carbon at temperatures in the 150 to 650°C (300 to 1200°F) range is more influenced by the precise composition (grade) of carbon than by the temperature.

A similar state of affairs exists for carbon at low temperatures. Hady et al.,[46] testing carbon seal grades submerged in liquid oxygen, found that performance improved as graphite content increased. Variation in wear rates among grades tested was greater than ten to one.

Temperature, of course, affects the viscosity of the fluid in the interfacial gap between a seal nose and its mating face. As the viscosity drops, the film becomes thinner, and asperity contacts become more frequent. Frictional work increases, and the seal nose temperature rises further above ambient. Considerable experience and judgment are required to estimate the running temperatures of the carbon itself. However, actual carbon temperature, rather than the environment temperature, is the controlling factor for both film-forming behavior and the chemical reactions of the carbon with contacting gases.

Figure 6.8 shows the effect of ambient temperature upon carbon wear rate. The rather complex test apparatus, like the one described by Brown[47] and Shevchenko and Brown,[40] simulates the main shaft seals of aircraft jet engines. Figure 6.9 is a sketch showing relationship of the components. Figure 6.10 is a drawing of the seal carbon, and Figure 6.11 is a photograph of the test head.

For the particular grade of carbon tested, P-4229, wear rate decreased with increasing temperature of the sealed gas up to 260°C (500°F). Above 260°C (500°F), wear rate increased slowly. The actual carbon temperature in this test apparatus is not known, but it is believed to be between the temperature of the sealed gas and that of the test chamber wall. Another indication of carbon temperature is the fact that coke

Fluid Sealed:	Heated Air
Fluid Pressure:	0.55 MN/m^2 (80 lb/in^2)
Fluid Temperature:	As Shown
Low Pressure Fluid:	MIL-L-7808C Oil Mist
Seal Nose:	P-4229
Mating Face:	Stellite 6B
Rubbing Speed:	56 m/sec (184 ft/sec)
Test Duration:	3 hours
Seal Chamber Wall Temperature:	As Noted

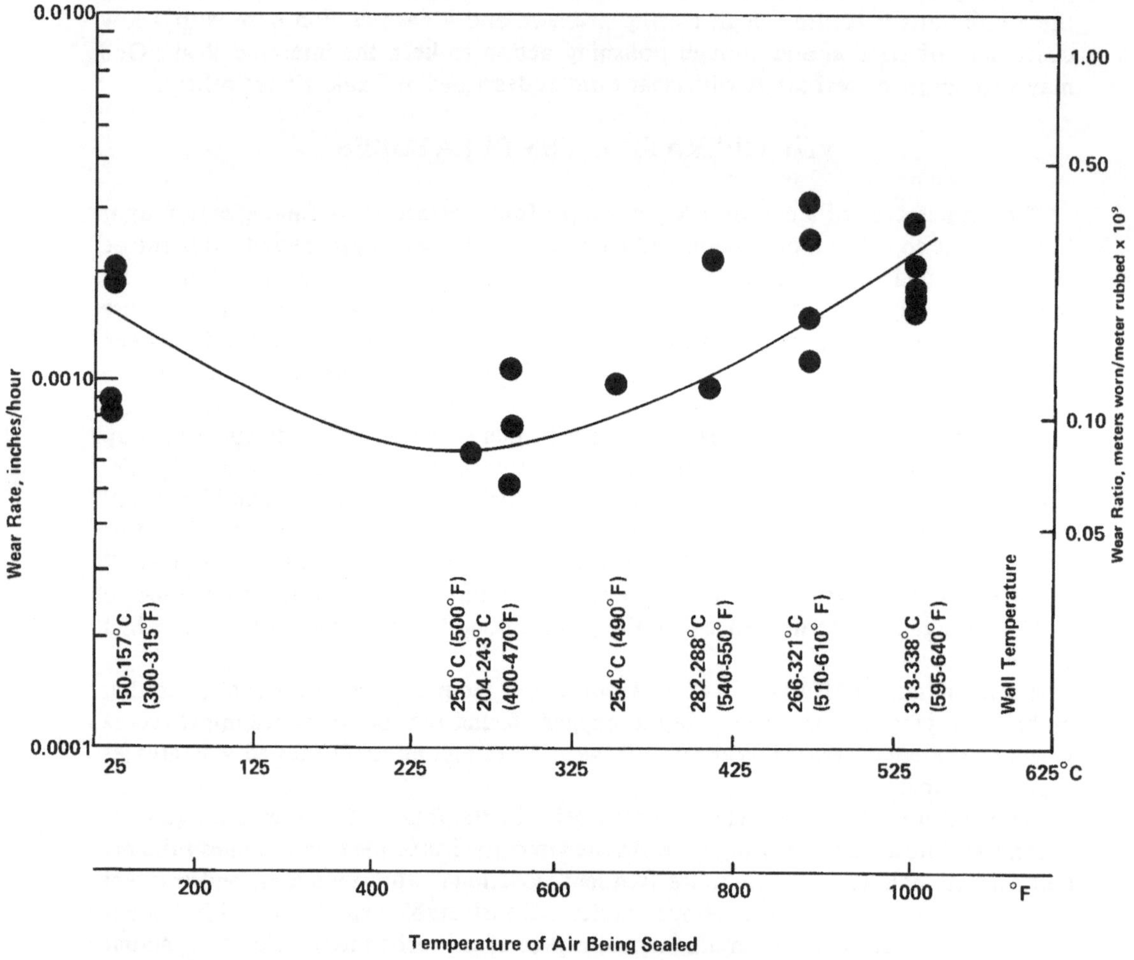

FIGURE 6.8. Air temperatures below 525°C have little influence on wear rate of P-4229 carbon seal.

and varnish formation, during a 3-hr test run, does not occur until the sealed gas temperature exceeds 425°C (800°F). Thermal degradation of the sealed oil is known to proceed at a significant rate above 375°C (710°F) in the presence of air.

Many other carbon grades have been tested in this apparatus. When sealing 38°C (100°F) air, variation in wear rates is relatively small (say, five to one). Above 260°C (500°F), however, some materials show wear rates at least 25 times greater than P-4229 — so fast, in fact, that they are completely destroyed in less than 1/2 hr of running. Thus, selection of the proper grade becomes progressively more critical as the operating temperature increases.

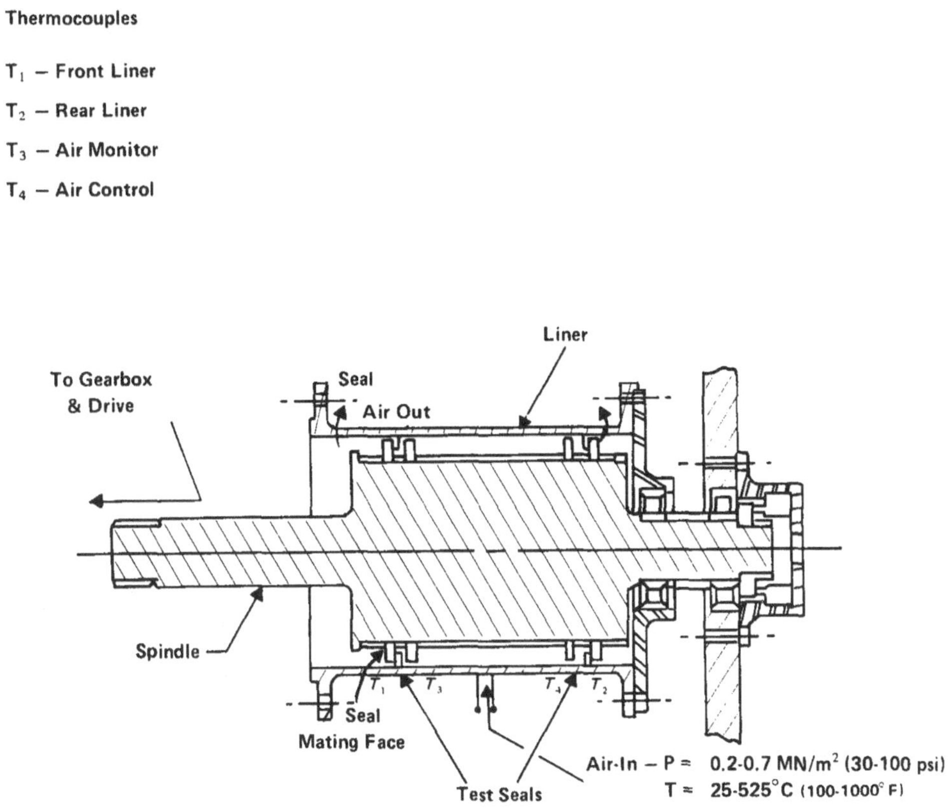

FIGURE 6.9. Test head for aircraft jet engine main shaft seals.

IX. RUBBING SPEED

Rubbing speed, coupled with the face pressure and coefficient of friction, dictates the rate of heat generation at the interface. There is considerable evidence to show that as speed increases, the shape of the hydrodynamic film pressure gradient across the seal face changes. As a result, wear rate and heat generation seldom rise in direct proportion to speed. Indeed, with sufficient hydrodynamic or areodynamic lift, seal wear can go down as rubbing speed increases. Figure 6.12 illustrates this effect with the seal and the test stand described in Figures 6.9 and 6.11. Leakage of air across the seal, coupled with oil mist cooling the back side of the mating face, probably kept the carbon temperature low enough to avoid oxidation.

Where these factors are not present, carbon temperature rises rapidly, and a point is reached where oxidation contributes significantly to wear. Figure 6.13, taken from Paxton and Shobert,[20] illustrates this. In this case, there were no hydrodynamic pads on the seal face, the pressure gradient across the face was smaller, and cooling by oil mist was less effective. At 85 m/sec (16,800 ft/min), bulk carbon temperatures had risen to 371°C (700°F), and the interface temperature was undoubtedly much higher.

Aircraft jet engines are, and will probably continue to be, the most demanding application for carbon seal materials. Each increase in engine efficiency brings with it higher seal temperatures, higher rubbing speeds, or both. McKibbin and Parks[48] describe some of the requirements while Dobek[49] and Fechter and Petrunich[50] describe some of the approaches being explored.

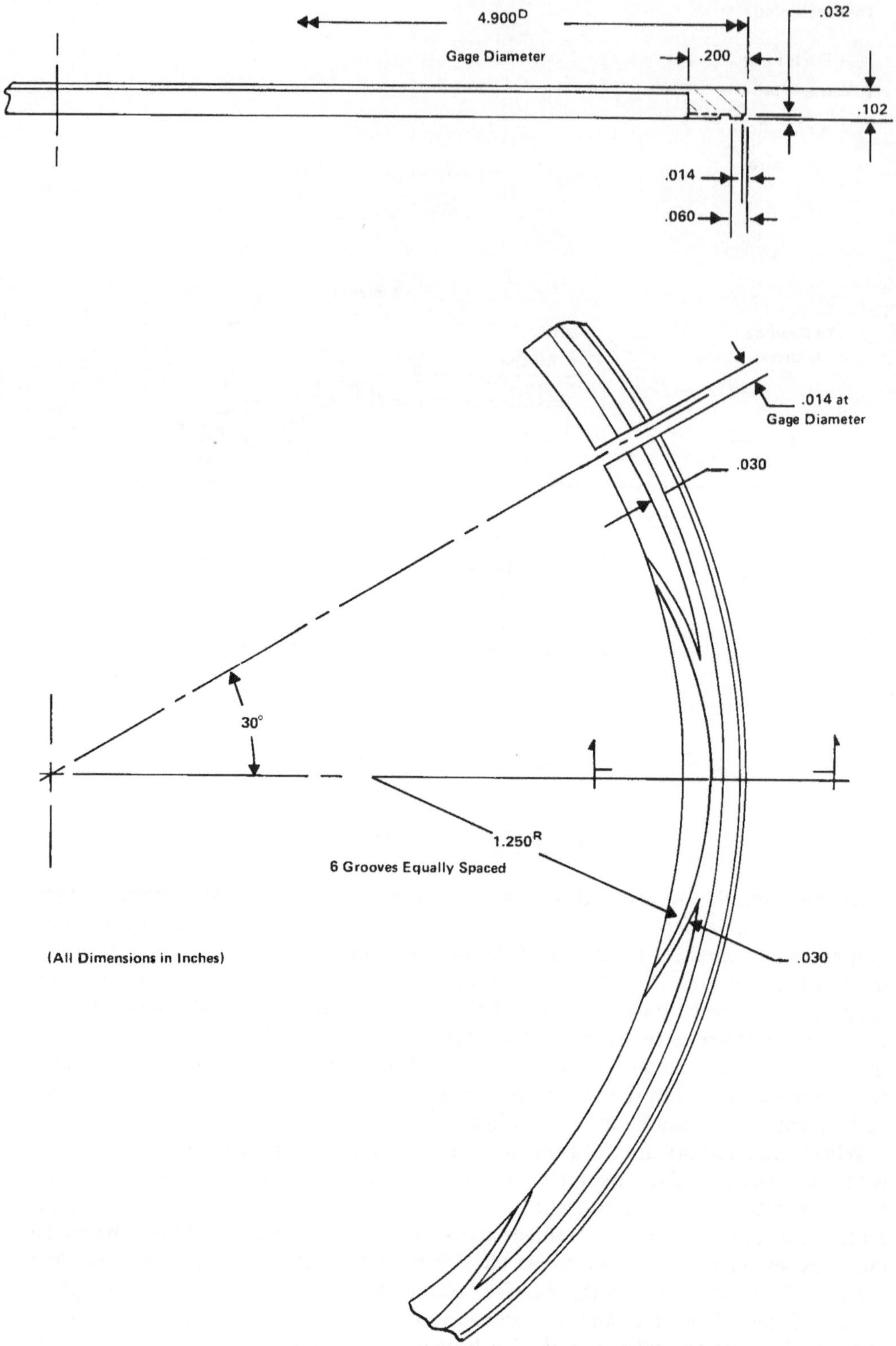

FIGURE 6.10. Seal ring for jet engine seal tester.

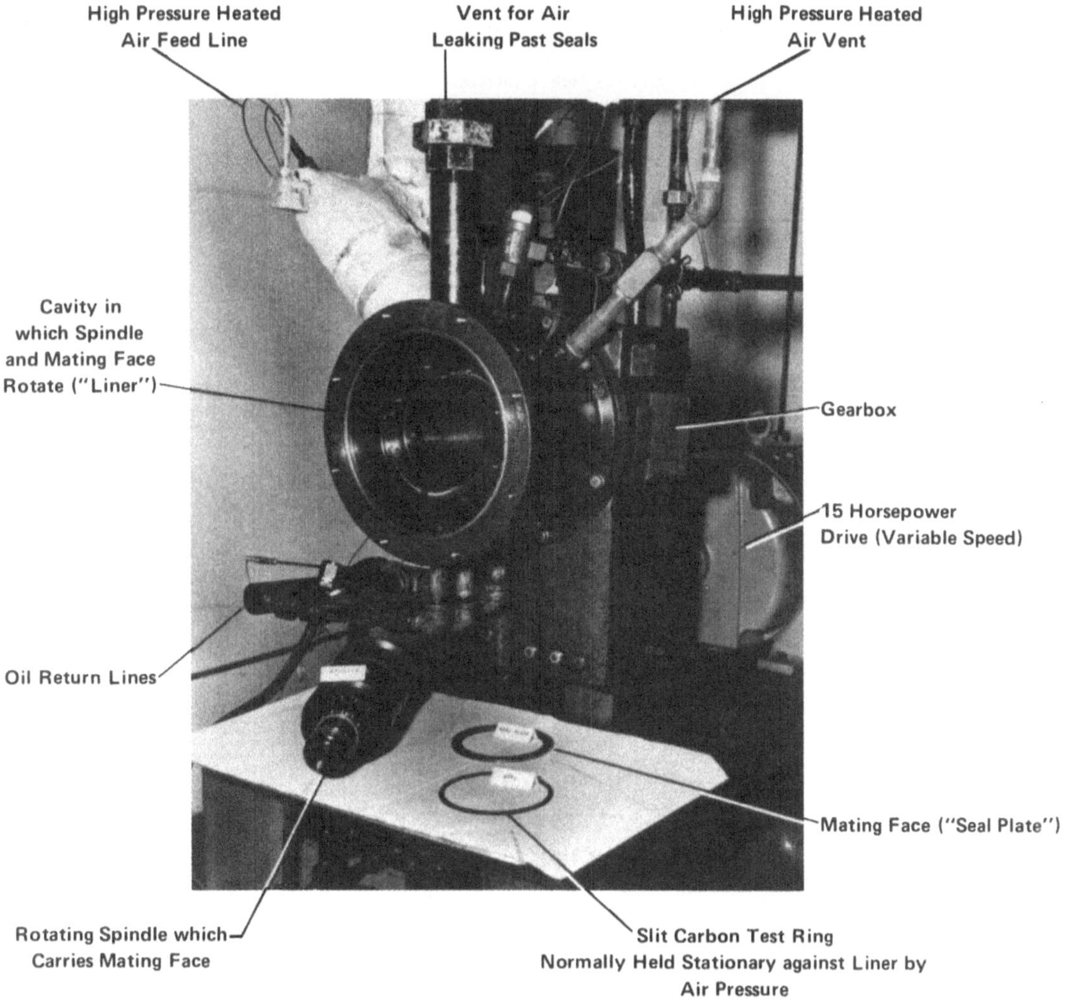

FIGURE 6.11. Jet engine seal tester.

X. PRESSURE GRADIENT ACROSS SEAL FACE

High pressure gradients apply large forces to the seal nose, tipping it enough to destroy the flatness of the lapped surface. As a result, the carbon nose has a line contact against the wear plate to carry the entire closing force. High wear occurs until the deformed seal nose conforms with the mating face surface. The process is repeated every time the pressure gradient changes. Mayer[51] and Snapp and Sasdelli[52] discuss the magnitude of these effects and present design strategies to minimize them. From a materials selection standpoint, all the designer can do is select a carbon grade with a high modulus of elasticity for applications where the pressure gradient varies from time to time.

XI. FACE PRESSURE

Carbon seal nose wear rates usually increase as the closing force applied to the face increases — usually, but not always. At very low wear ratios, say below 10^{-13} m worn

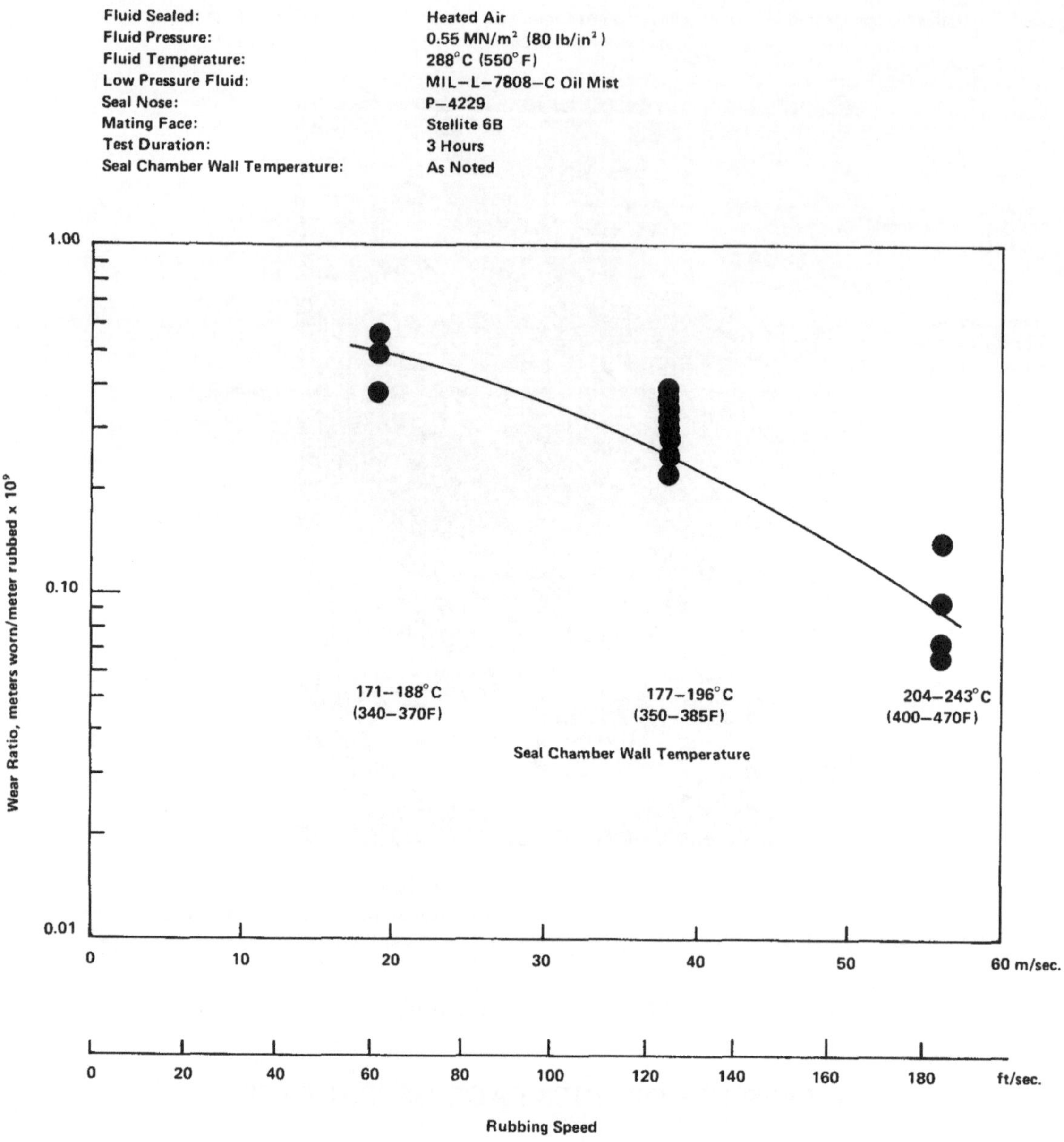

FIGURE 6.12. Wear ratio of carbon seal nose drops with increasing rubbing speed when mating face is kept cool.

per meter rubbed, wear is insensitive to face pressure. The faces are separated by a hydrodynamic film 99.9% of the time. Solid-solid contact occurs only during start-up and operating transients.

A good example of this operating regime can be found in the face seals used in automotive air conditioning compressors.[53] The refrigerant in these systems contains a small amount of oil. Some of this oil migrates to the rubbing faces of the shaft seal on the refrigerant compressor. This film of oil forms a hydrodynamic film separating the faces when the compressor is operating even though the closing forces can reach 1 MN/m^2. The wear ratio of the carbon nose is usually well below 10^{-13}. It is not unusual for a 35-mm diameter seal nose to show less than 50 μm wear after 3000 hr at an average shaft speed of 1000 rpm. Probably most of this occurred during starts and stops.

Fluid Sealed:	Air
Fluid Pressure:	0.03 MN/m² (4 lb/in²)
Fluid Temperature:	39°C (100°F)
Low Pressure Fluid:	MIL–L–7808C Oil Mist
Face Pressure:	0.08 MN/m² (12 lb/in²)
Mating Face:	Chrome Plated AISI 4140
Test Duration:	6 hours

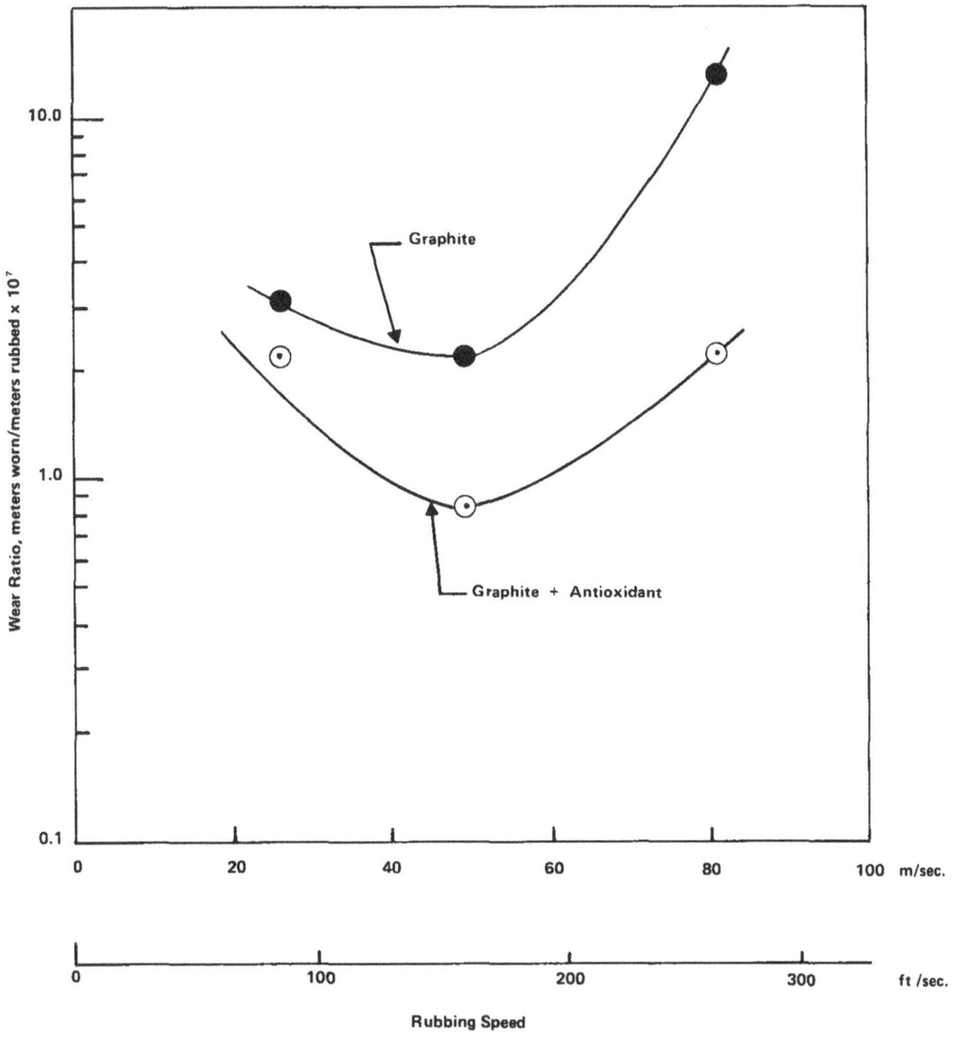

FIGURE 6.13. Wear ratio of carbon seal nose drops with increasing speed until nose overheats. (From Paxton, R. R. and Shobert, W. R., *ASLE Trans.*, 5, 308, 1962.)

The middle part of the wear vs. nose pressure relationship is characterized by wear ratios ranging from 10^{-12} to 10^{-10} m per meter rubbed. In this zone, wear rates clearly increase as face pressure increases, but the relationship is hard to define with any precision. Replicate tests, even continuous constant speed tests lasting 100 hr, show wide variations in wear rate. Figure 6.14 illustrates repeatability for what are essentially 19 replicate runs. In this illustration a carbon nose, grade P-658-RC, was run against an alumina wear plate, grade AD-85. The fluid sealed was clear, warm water, the rubbing speed was constant, and the face pressure was held at 0.40 ± 0.03 MN/m². Despite this attempt to control running conditions during a 100-hr test, wear ratio ranged from

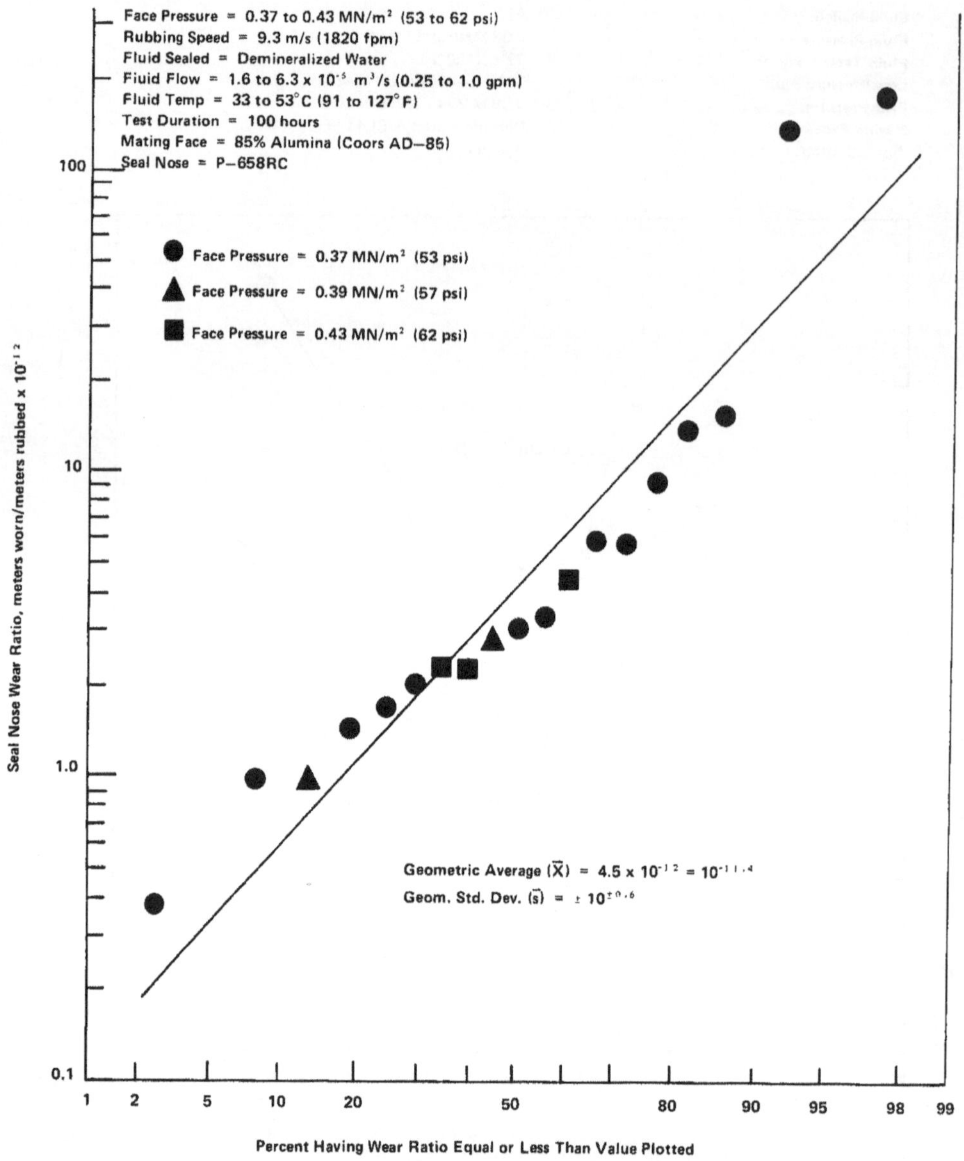

FIGURE 6.14. Variability of replicate wear tests.

0.0004×10^{-9} to 0.17×10^{-9} metre worn per metre rubbed. The data show a Gaussian probability distribution with a geometric standard deviation of $\sim 10^{\pm 0.6}$ m/m. These variations in wear rate are believed due to loss of the hydrodynamic film for 1 to 40% of the test duration.

The third portion of the wear rate vs. pressure relationship shows wear ratios generally above 10^{-10} m/m, too high for most face seal applications. Solid-solid contact prevails and the role of the fluid is that of a coolant. Wear rate depends very much upon nose and mating face materials. Differences as great as 1000-fold would not be unusual. For any given pair and fluid environment, however, wear ratio is related to contact pressure and is quite reproducible. Wear data on over 50 material pairs at contact pressures ranging from 0.1 to 6 MN/m² are correlated by

$$\text{Wear Ratio (dimensionless)} = kP^{0.8}$$

where k depends on rubbing pair, temperature, and units of face pressure. This relationship eventually "falls apart" when the temperature of the interface becomes high enough to alter the physical properties of one of the rubbing surfaces.

Face pressure has a greater effect on seal life than any other factor under the designer's control once he has selected the sliding materials. Wherever possible, spring pressure to close the seal should be kept below 0.15 MN/m^2 (21 lb/in.^{2}). System pressure adds to this spring pressure an amount which is dependent upon the seal design. Rubbing velocity reduces face pressure by an amount dependent on hydrodynamic or aerodynamic lift. Table 6.2 shows approximate face pressure limits based on the best carbon mating-face couples available in the mid-1970s and assumes the wear ratio must be below 10^{-9} m worn per meter rubbed.

XII. LEAK RATE

A refrigeration compressor seal has a maximum permissible leak rate of 10^{-6} cm^3/min.[51] At the other extreme, large aircraft jet engine shaft seals can tolerate a leakage of 0.3 m^3/min (12 ft^3/min).[36] A "no leak" face seal for an aircraft fuel pump means, in practice, a leakage of 0.2 to 2 cm^3/hr,[51] while face seals are considered acceptable for general purpose usage if leakage remains below 200 cm^3/hr.[55] The definition of what is an acceptable leak rate, as these examples show, depends on the intended usage of the seal.

Four leakage paths exist for face seals: (1) across the rubbing interface, (2) across the axially sliding secondary seals, (3) beneath 0 rings or other static seals, or (4) through the body of the seal nose and its mating face. Body leakage of porous carbon seal materials is reduced by impregnation, as described earlier. Some impregnants are relatively ineffective in preventing leakage, and as a consequence, seal carbons in commercial use vary seven orders of magnitude in their permeability. Where low leakage is required in the face of high-pressure gradients and low-fluid viscosities, the designer must take pains to specify low-permeability seal and mating face elements.

Low-permeability seal materials are generally more expensive and occasionally exhibit higher friction coefficients caused by the impregnant. Because of these two factors, the designer should not require low-permeability materials unless the application at hand demands it.

XIII. SEAL LIFE EXPECTANCY

Stationary gas turbines require 15,000 hr of operating life while the drive turbine in a rocket fuel pump or torpedo engine has a life measured in minutes. Between these

TABLE 6.2

Face Pressure Limits for Seals

Typical fluid at interface	Viscosity range, centipoise (Ns/m^2)	Temperature range °C(°F)	Face pressure limit MN/m^2 (lb/in.^{2})
Air methane, steam, carbon dioxide, hydrogen, etc.	0.005—0.05 (0.5—5 × 10^{-5})	0—538 (40—1000)	0.15 (21)
Water, gasoline, methanol, kerosene	0.1—5 (1—50 × 10^{-4})	0—316 (40—600)	2.0 (290)
Light oils	5—50 (5—50 × 10^{-3})	0—316 (40—600)	4.0 (580)

extremes lie the majority of seal life requirements. A seal that lasts over 1000 hr is generally considered acceptable for most liquid-handling pumps.[55,56] Longer life is required for seals on consumer items such as automotive water pumps, air conditioning refrigeration compressors, or home heating circulating water pumps.

Seal life is often limited by wear of the rubbing members. This being so, careful selection of seal and mating face material is a critical element in the overall design. Table 6.3 is a general guide for selecting carbon nose materials. It is no substitute for experience; rather it is a "first aid" to be relied upon only until the designer can obtain advice from specialists in the field.

XIV. SIZE OR CONFIGURATION

Probably 99% of all carbon seals are between 25 and 150 mm (1 and 6 in.) in diameter, and most materials are available over this size range. As he moves into larger sizes, however, the designer must accept a smaller selection of materials. Even where pressing facilities and impregnating autoclaves are big enough to produce the grade desired, poor yields during baking of the larger shapes lead to a sharply higher manufacturing cost and, hence, higher price.

XV. DESIGN OF CARBON SEAL COMPONENTS

A careful designer for seal components considers the performance (and life) required, the cost of assembly, and the cost of manufacture of both the carbon and the associated components. In the U.S., tentative standards have been set by the ASLE[57] to provide greater interchangeability of seals and seal components.

A. Molded-to-Size Carbon Seal Noses

Whenever the usage exceeds 1000 parts per year, the designer should consider the possibility of molding the configuration required in the seal nose. Molding techniques for carbon are similar to those used for powdered metal parts. Carbon powder will not "flow" laterally; hence, all configurations must be generated by direct compression. Some limitations are

1. No negative drafts or reverse sections. Except for the outer diameter, all surfaces normal to the direction of molding force, such as the sides of shoulders or counterbores, require a draft of 1 to 3° to permit release from the punch face.
2. Close tolerances must be avoided. Carbons shrink during baking, and this shrinkage is not completely predictable. Depending upon the carbon grade, a tolerance ranging from 1 to 4% (± 0.5 to ± 2%) will be required in dimensions normal to the molding direction. Tolerances parallel to the molding direction must be up to 50% greater due to the added difficulty of controlling die fill when using carbon molding powder.
3. Shoulder or counterbore height should not exceed 1/3 the overall height of the piece, and shoulder radial thickness should be at least as great at its height.
4. Inside corners should have a radius (fillet) of 0.1 to 0.3 mm (0.005 to 0.015 in.).
5. Outside corners should have a minimum chamfer of 0.2 mm (0.010 in.) × 45°.
6. A seal face in cantilever with reference to the skirt should be avoided.

Figure 6.15 illustrates these recommendations for molded-to-size parts, and Figure 6.16 shows some typical molded-to-size seal rings. In each of these applications the only machining operation required is lapping to generate the flat, smooth sealing surface. Figure 6.17 shows examples where most of the configurations are molded and where one or two machining operations are needed to complete the seal ring.

TABLE 6.3

Selector Chart for Carbon Nose Materials

	Fluid sealed					
			Aqueous solutions		Oil, solvents	
Temperature	Dry gas	Humid air, exhaust gas	Clear	Dissolved solids	Below 10 cp	Above 10 cp
−273 to −35°C (−460 to −30°F)	Medium to high graphite level plus metal salt additive		—	—	Low to medium graphite level plus metal salt additive with mild polishing action	
−35 to 125°C (−30 to 260°F)	Same as above plus polishing	High graphite level unless PV below 10,000 lb ft/in.^{2} min	Low to medium graphite level	Low graphite level, medium polishing	Low to medium graphite level	Medium graphite level, very fine texture, low porosity
125 to 275°C (260 to 530°F)	High graphite with friction-reducing additive		Low to medium graphite plus metal impregnation	Low graphite plus polishing	Low to medium graphite	Low to medium graphite, fine texture, polishing
275 to 425°C (530 to 800°F)	Electrographites with friction-reducing additive		Low graphite plus metal impregnation	Low graphite plus polishing	Low graphite plus heavy polishing to control coke	
425 to 650°C (800 to 1205°F)	Hard electrographite with antioxidant and friction-reducing additives		—	—	—	—

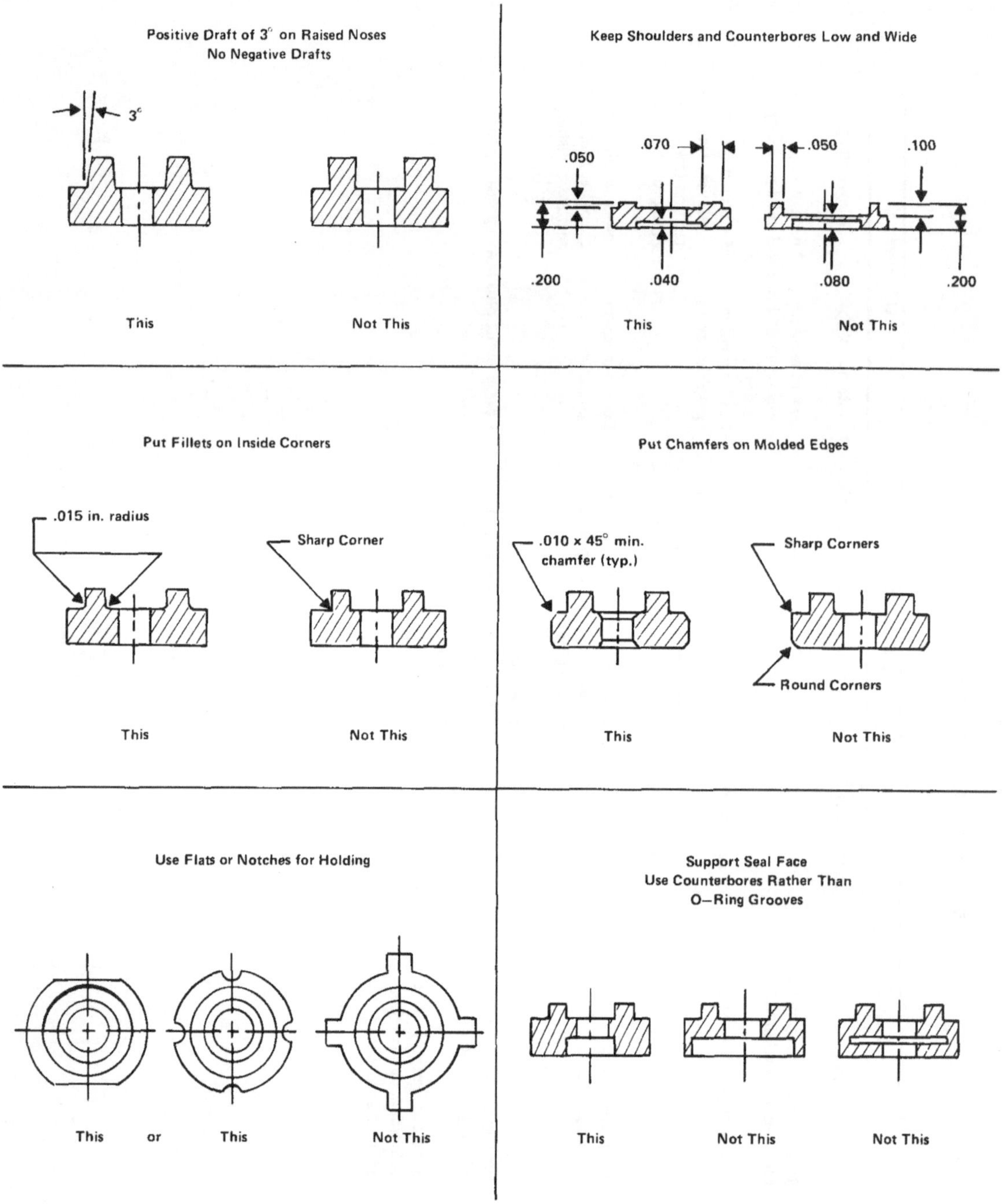

FIGURE 6.15. Recommendations for molded-to-size parts (all dimensions in inches).

B. Machined Seal Noses

More often than not the tolerances required for molding are more than the designer can allow. In these cases attention to the following suggestions will help to keep the cost low:

1. Notches (holding slots) or flats are less costly than protruding ears on the outside diameter.

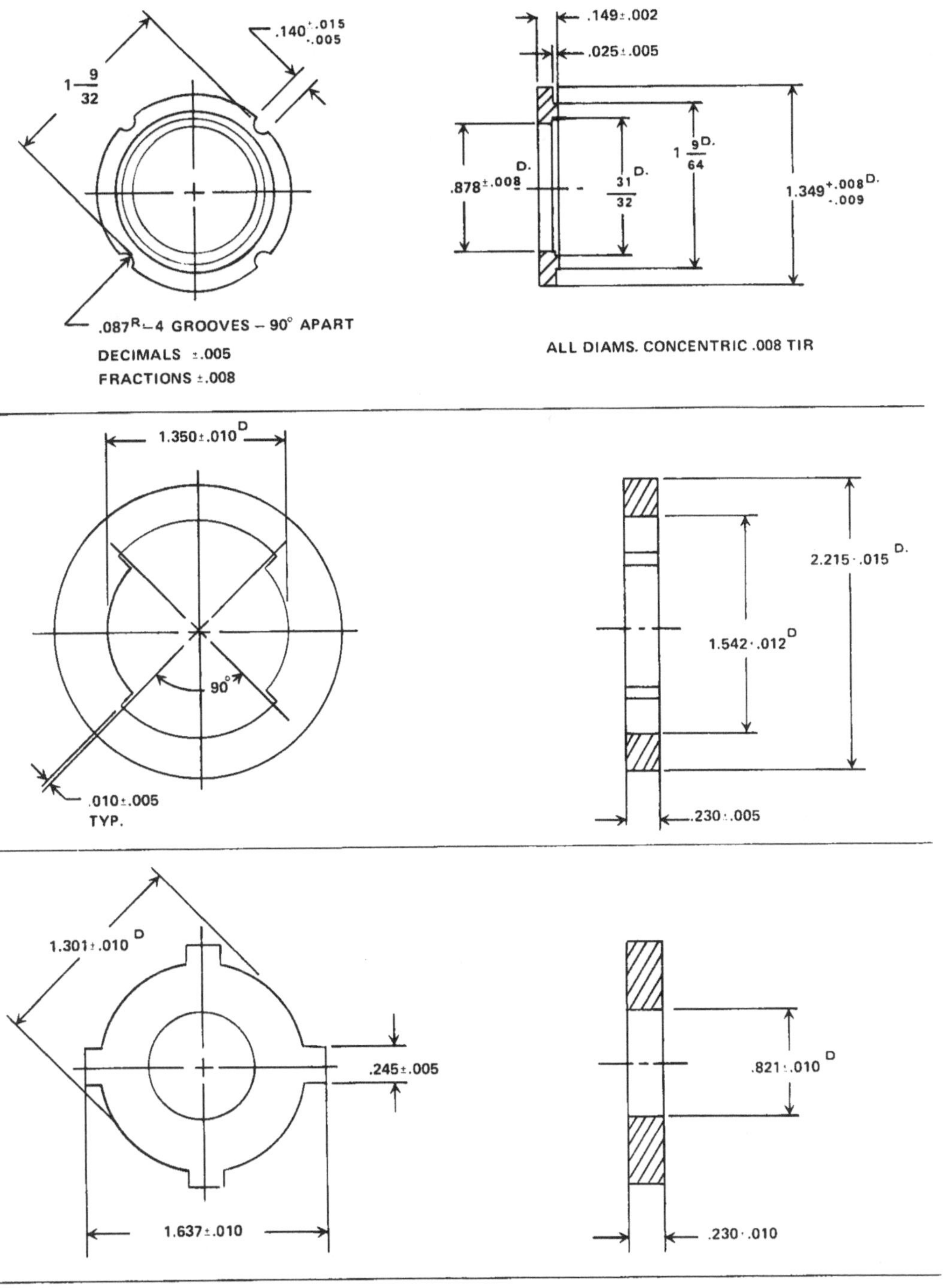

FIGURE 6.16. Typical molded-to-size carbon seal noses (all dimensions in inches).

2. Notches are less costly on the outside diameter than the inside diameter.
3. Chamfers on external corners are cheaper than rounds.
4. Counterbores are cheaper than O-ring grooves.
5. The tolerances shown in Table 6.4 are "standard"; cost climbs rapidly for tighter tolerances.
6. A "commercial lap" will give flatness to 0.005 mm (0.0002 in.). For most appli-

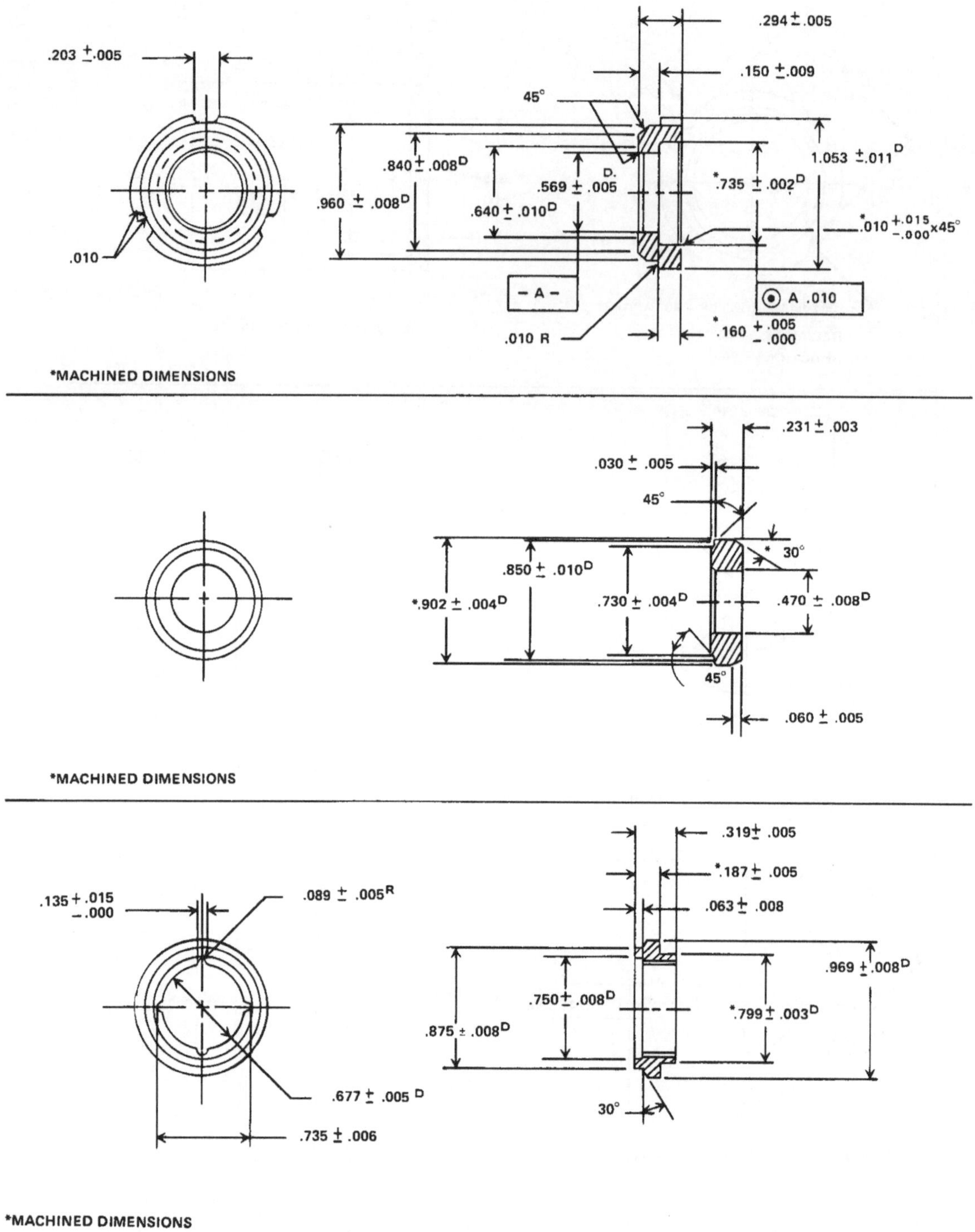

FIGURE 6.17. Typical partially molded carbon seal noses (all dimensions in inches).

cations sealing liquid of this flatness is good enough. After a few minutes run in, the carbon will give a tight seal.

7. Flatness to 2 to 3 light bands (monochromatic helium light) per inch (0.3 μm/cm) is required where the seal must be gastight when first assembled. One light band (He) is 11.8×10^{-6} in. (0.3 μm).

Figure 6.18 shows several machined face seal rings while Figure 6.19 shows a variety of seal rings. The carbon components of ring seals can become very intricate as this photograph shows.

XVI. IMPERVIOUS TESTING

Manufactured carbon ps inherently a permeable material having interconnected porosity ranging from 6 to 30%. To convert this material into a refrigeration seal nose with a vapor transmission below 1/2 oz of Freon per year requires close attention to manufacturing and design parameters. A wall thickness of 1.6 mm (0.063 in.) or 8% of the seal shoulder diameter, whichever is greater, is the first requirement. The impregnant used to seal the interconnected pore structure must be fluid enough to penetrate the entire structure and, of course, must be unaffected by the Freon or other gas being sealed. After impregnation, stock removal from any surface must not exceed 0.5 mm (0.020 in.) if one is to avoid the risk of opening up initially closed — and, hence, unimpregnated — porosity.

Finished seals are usually tested for imperviousness by clamping the faces between gaskets and pressuring the internal cavity with dry nitrogen gas. Pressures to 1.4 MN/m^2 (200 psi) may be used providing the carbon wall is thick enough to tolerate the hoop stresses generated. While the ring is thus pressurized, the outside is painted with a nonstaining liquid surfactant and then scanned for bubbles. Cracked rings and "leakers" are quickly spotted and discarded. Experience has shown that 99% of the rings which pass this test will also pass a mass spectrometer-helium leak test or a halogen leak test.

Before any leak test, seals must be thoroughly dried. Liquid water blocks open pores and, for quite some time, conceals the fact that the seal body is permeable to gases.

If the seal is intended for liquid use, a gas pressure test of 0.17 MN/m^2 (25 psi) is ample to assure a level of permeability low enough for most applications. Specifying a higher test pressure adds to the costs and, more important, greatly circumscribes the variety of materials that can be used.

XVII. INSPECTION OF SEAL CARBONS

The major carbon producers use statistical quality control methods coupled with in-process inspection at every level of manufacture from receipt of the raw material to final lapping of the finished part. Before shipping, each production lot is again sampled and inspected. The most common sampling program is that set forth in MIL Std. 105-D,[58] and the standard average quality level (AQL) is 2.5%. Tighter standards can usually be obtained for a slight premium.

XVIII. FAILURE ANALYSIS

When a seal fails to perform as expected, the carbon component is usually the first suspect. Several complaints and common findings are discussed in the following paragraphs.

A. Seal Leaks When First Installed or Early in Its Life

The first thing to establish is where the leakage occurs — at the interface, through the carbon body, or at the O ring (or secondary seal element). Where there is face leakage when the seal is at rest, check the carbon and mating face for flatness and parallelism. Check also for dirt or grit between the faces. Flatness at room temperature does not guarantee flatness at operating conditions. Thermal or pressure distortion

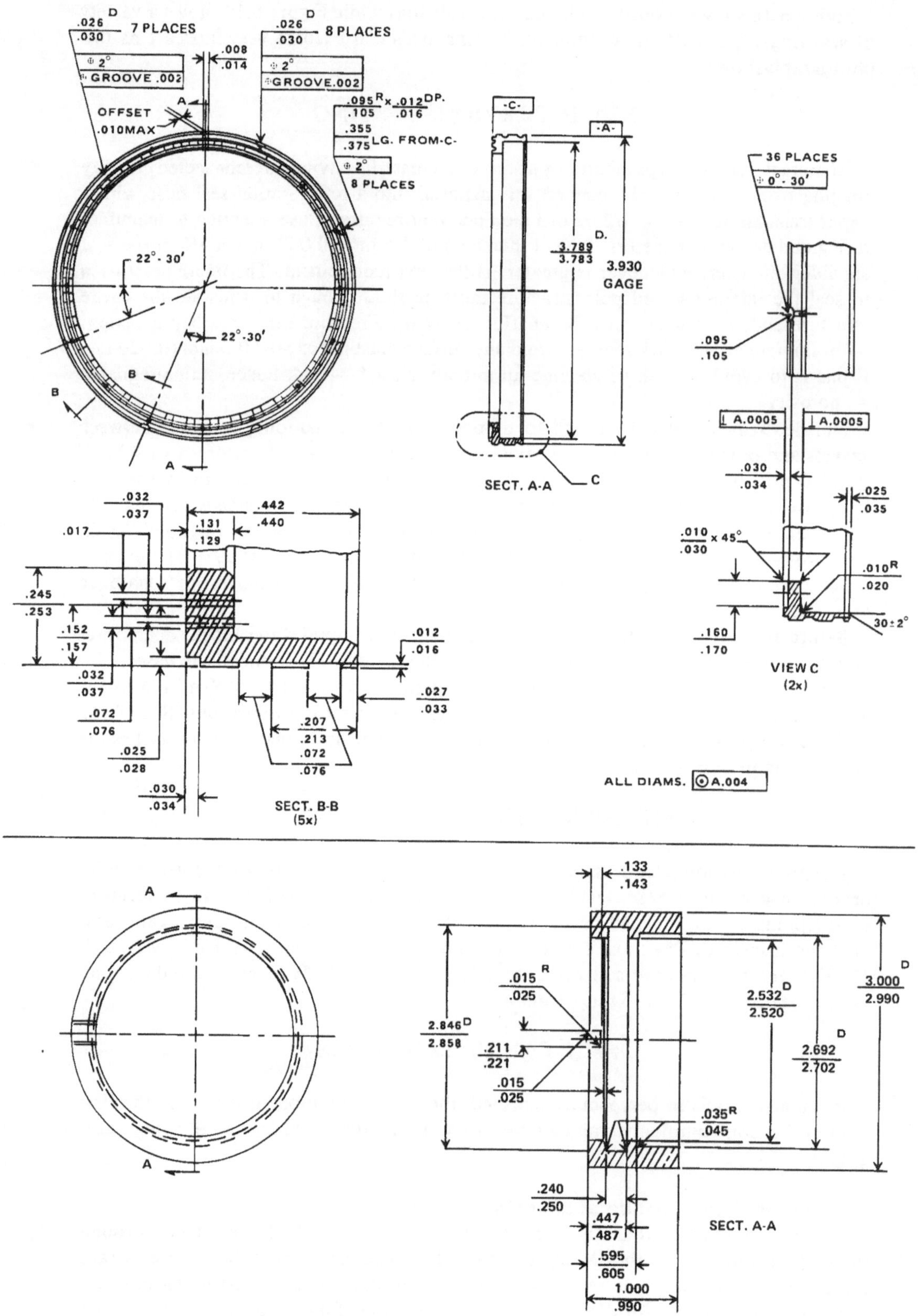

FIGURE 6.18. Typical machined carbon seal noses (all dimensions in inches).

TABLE 6.4

Normal Tolerances For Carbon Seal Nose Rings

	Tolerances[a]			
	Machining		Molding[b]	
Surface produced by	mm	in.	mm	in.
All diameters	0.10	0.004	0.25	0.010
Thickness from lapped seal face	0.25	0.010	0.25	0.010
Thickness between other faces	0.25	0.010	0.5	0.020
Other linear dimensions	0.25	0.010	0.25	0.010
Flatness of lapped face, TIR	0.005	0.0002	—	—
Parallelism of flat faces, TIR	0.10	0.004	0.25	0.010
Squareness of diameters to seal face, TIR	0.10	0.004	0.10	0.004
Concentricity of diameters, TIR				
Seal diameter to 25 mm (1 in.)	0.10	0.004	0.12	0.005
Seal diameter 25—37 mm (1—1.449 in.)	0.10	0.004	0.2	0.008
Seal diameter 37—51 mm (1.5—1.99 in.)	0.10	0.004	0.3	0.012
Seal diameter 51—75 mm (2.0—2.999 in.)	0.10	0.004	0.4	0.016
All angles	2°		2°	

[a] Total spread from minimum to maximum value. Molding tolerance must be used if either surface used for measurement is molded.

[b] Or 2% of stated dimension, whichever is greater.

may occur in either rubbing member, though the carbon is more vulnerable because of its low modulus. Another cause is inadequate closing force from incorrect springs, swollen O rings, high sliding friction in the O ring, misalignment, or overcompensation.

B. Seal Leaks Only When Running

If the face leak occurs only when the seal is running, misalignment or chatter (vibration) is the probable cause. A high coefficient of friction tends to make chattering worse. The dynamics of seal chatter are presented in mathematical terms by Chiang and Cheng.[59] Hudelson[60] ran tests on an undamped bellows seal where wear rates — or, more properly, mechanical erosion rates — of 0.18 mm/hr (0.007 in./hr) were caused by chatter. Vibration frequencies were 3 to 6 kHz, and accelerations exceeded 100 g's, with the predominant mode being torsional and rocking. No carbon seal nose can bear up under this kind of punishment.

Leakage through the body usually indicates a cracked carbon. Excessive stock removal after impregnation can also open up a leak path in a carbon ring which was initially gastight. One should also check the carbon grade used to be sure it is an impervious grade.

Scratches in the counterbore or a rough machining job are common causes for leakage at the O ring. A less common cause is the destruction of the O ring by overheating or by chemical attack.

C. Leaks in Circumferential Shaft Seals

Radially acting shaft seals present special problems not present with face seals. Most of these assemblies contain a number of carbon elements that are spring-loaded to maintain contact with themselves and the shaft.

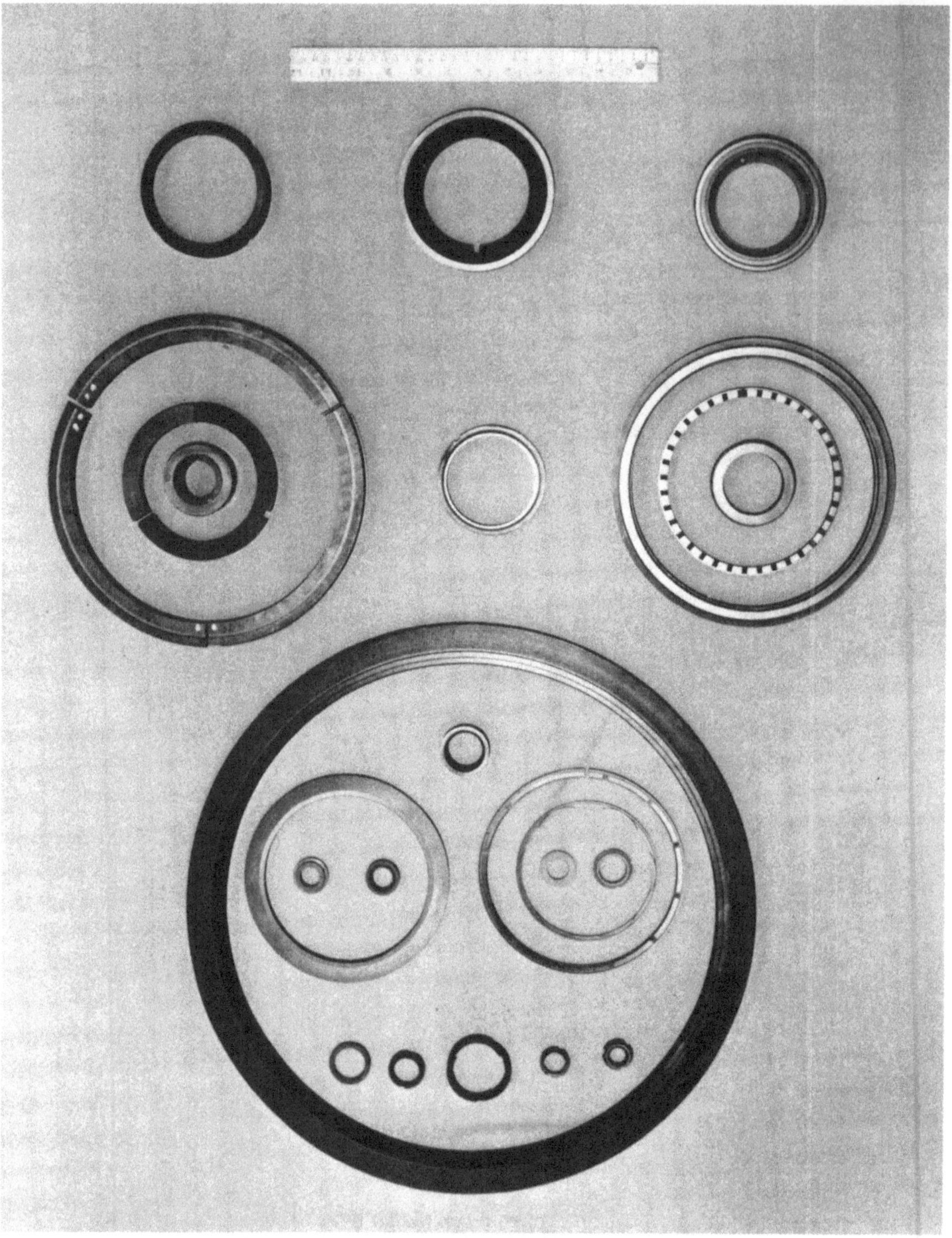

FIGURE 6.19. Carbon seal noses.

Before a seal of this type is installed, it should be carefully examined to be sure there are no broken carbons and that the carbon elements are free to move in the radial direction. Individual elements should be checked for markings indicating match-mated components. Elements from several seal assemblies should never be mixed. High leakage rates will occur if elements are not assembled exactly as specified.

Frozen or "hung-up" carbons in circumferential seal assemblies may be caused by any of the following:

1. Improper machining or assembly of carbons
2. Carbons that stick together because of contamination by coolants or grease picked up during machining or assembly
3. Carbons that stick together due to bleeding out of the impregnant — this may be caused by overheating the carbon or, more likely, by allowing the carbon to be wet for prolonged periods resulting in hydrolysis and bleeding out of the impregnant. (Many high temperature impregnants are slowly hydrolyzed by water.)
4. Carbons that stick to the metal housing — Galvanic corrosion of the metal is the usual cause of this. It may occur if the seal assembly is allowed to stand in a wet condition. Most high-temperature carbon materials contain inorganic salts which will supply ions in the presence of water. Even carbons completely free of inorganic materials may have been contaminated with soluble minerals during machining. Dissimilar metals in the springs and housing of the assembly supply the electromotive driving force, thereby energizing the galvanic couple.

Corrosion of shaft seals seldom occurs after installation due to the presence of oil and the usual precautions taken to protect the entire piece of equipment while in storage. The damage usually occurs to the seal assembly before it is installed. Radial seal assemblies should be thoroughly dried and then sealed in vaporproof bags. If removed from these bags for inspection, they should be returned to the bag and the bag resealed until the assembly is to be installed.

D. Seal Performs Adequately but Fails Prematurely

A number of factors can force replacement of a seal short of its normal service life. Only those factors related to the carbon and its mating face will be discussed below.

Cracks — When disassembly of the seal shows the carbon or the mating face to be cracked, one of the following conditions is usually the cause. Thermal shock is the usual reason for cracked ceramic faces. The seal may have run dry for a while and become very hot. Then liquid was suddenly readmitted, chilling the ceramic surface and generating intolerable thermal stresses. Carbon rarely fails from thermal shock. A cracked carbon, assuming it was intact when installed, generally results from a mechanical or hydraulic overload.

High carbon wear — The most common cause of high carbon wear is use of the wrong carbon grade in the first place. The next most common cause is overloading due to misalignment, improper installation, or use at above-design pressures. Overheating due to inadequate cooling of the seal chamber and the presence of suspended or precipitating solids in the seal chamber will also accelerate carbon wear. Overheating of the rubbing elements may also be caused by a design which fails to provide an effective path for heat flow from the rubbing surface.

High leakage — When disassembly shows no cracks and no excessive wear, the most probable cause is hang-up of the assembly by solids. Less common causes of leakage are corrosion of the spring providing the closing force, or corrosion of the carbon rendering it permeable, or corrosion of the metal bellows.

E. Seal Develops High Leakage Early in Life

The usual cause of excessive leakage early in seal life is the inability of the spring to close the seal due to a build-up of solids which restrict its axial movement. Less common is swelling of the O ring or development of "blisters" on the carbon face. Strugala[61] has shown that cycling temperatures will develop blisters in seal faces even at room temperatures. Blistering (or spalling) of the seal face can also occur if the seal is rapidly heated to a temperature high enough to cause vaporization of the resin impregnant.

XIX. SUMMARY

The design and construction of mechanical seals is a highly specialized and rapidly advancing art. In the U.S. alone there are a score of firms in this field. Close cooperation between these firms and the carbon producers has been one of the reasons for the rapid improvement in performance and reliability of mechanical seals.

The seal manufacturer, however, is completely dependent upon his customer for information on operating conditions. This customer, typically a pump manufacturer, obtains his information from the process designer. Thus, any information regarding operating conditions has passed through three agencies before it reaches the carbon supplier. Added to the usual errors and distortion inherent in such an information transfer, one has the understandable reluctance of the process engineer to disclose what he considers proprietary information regarding the fluids to be handled.

Standing at the end of this tortuous information pipeline is the carbon supplier who is asked to select a grade which will give suitable performance. It is a tribute to the engineering skill of all parties that the system works as well as it does.

Certainly progress in developing improved rubbing elements for mechanical seals can be made only when factual information on performance and operating conditions reaches the supplier of those elements. My hope is that the information of this chapter will promote a better understanding of the role of the manufacturer of rubbing elements and will encourage better information transfer among all parties using his products.

REFERENCES

1. **King, A. L.**, *Bibliography of Fluid Sealing*, British Hydromechanics Research Association, Essex, England, 1962.
2. **Stair, W. K.**, A Bibliography on Dynamic Shaft Seals, No. ME-5-62-3, University of Tennessee, Knoxville, 1962.
3. **Ewbank, W. J.**, Dynamic Seals — A Review of the Recent Literature, Paper 67-WA/Lub.-24, in American Society of Mechanical Engineers, Winter Annual Meeting, Pittsburgh, Pa., November 12 to 17, 1967.
4. **Elwell, R. C. et al.**, Study of Dynamic and Static Seals for Liquid Rocket Engines, General Electric Co. under NASA contract NAS7-102 (available to U.S. government agencies and U.S. government contractors only.)
5. **Koenig, T. H.** Review of dynamic seal literature through 1967, *ASLE Special Publication SP-2*, American Society of Lubrication Engineers, Park Ridge, Ill., 1969, 37.
6. **Mayer, E.**, *Mechanical Seals*, Iliffe Books Ltd., London, 1969.
7. **Schoenherr, K.**, Design Terminology for Mechanical End Face Seals, SAE Paper 650301, Society of Automotive Engineers, Warrendale, Pa., 1965.
8. **Brown, P. F., Ed.**, A glossary of seal terms, ASLE Special Publication SP-1, American Society of Lubrication Engineers, Park Ridge, Ill., 1969.

9. **Dahlheimer, J. C.,** *Mechanical Face Seal Handbook,* Chilton Book Company, Philadelphia, 1972.
10. **Cawley, L. J. and Duncan, R. M.,** High performance seals, *Mach. Des.,* 43, 152, 1971.
11. **Giles, O.,** New design concepts and materials for mechanical shaft seals, *Mech. Eng.,* 94, 18, 1972.
12. **Wilkes, G., III,** Analysis and development of a dry face mechanical shaft seal concept, *Lubr. Eng.,* 26, 273, 1970.
13. **Hershey, L. E.,** Time rate of wear testing mechanical seal carbons, *Lubr. Eng.,* 24, 27, 1968.
14. **Paxton, R. R. and Shobert, W. R.,** Testing carbons for seals and bearings, *Lubr. Eng.,* 17, 27, 1961.
15. **Abar, J. W.,** Rubbing contact materials for face type mechanical seals, *J. Am. Soc. Lubr. Eng.,* 20, 381, 1964.
16. **Savage, R. H.,** Graphite lubrication, *J. Appl. Phys.,* 19, 1, 1948.
17. **Smith, R. N., Pierce, C., and Joel, C. D.,** The low temperature reaction of water with carbon, *J. Phys. Chem.,* 58, 298, 1954.
18. **Snow, C. E., Wallace, D. R., Lyon, L. C., and Crocker, G. R.,** Reaction of carbon blacks with oxygen, in *Proc. 3 Conf. Carbon,* Pergamon Press, Oxford, 1957, 279.
19. **Buckley, D. H.,** Friction, Wear and Lubrication in Vacuum, NASA SP-277, National Aeronautics and Space Administration, Washington, D.C., 1971.
20. **Paxton, R. R. and Shobert, W. R.,** Testing high speed seal carbons, *ASLE Trans.,* 5, 308, 1962.
21. **Mayer, E.,** Mechanical Seals, *Iliffe Books Ltd., London,* 1969, 104.
22. **Janetz, R. W.,** Mechanical seals in pumps for hydronic systems, *Lubr. Eng.,* 19, 321, 1963; *Lubr. Eng.,* 21, 372, 1965.
23. **Piehn, L. D.,** The application of tungsten carbide for mechanical face seals, *Lubr. Eng.,* 21, 381, 1965.
24. **Voitik, R. M.,** A new approach to realistic wear testing, *J. Am. Soc. Lubr. Eng.,* 19, 490, 1963.
25. **Paxton, R. R. and Strugala, E. W.,** Wear rates of impregnated carbons sealing warm water, *Lubr. Eng.,* 29, 341, 1973.
26. **Iny, E. H.,** Theory of sealing with radial face seals, *Wear,* 18, 51, 1971.
27. **Kojabashian, C. and Richardson, H. H.,** A Micropad Model for the Hydrodynamic Performance of Carbon Face Seals, Paper E-4, in *3rd Int. Conf. Fluid Sealing,* Cambridge, England, April 1967.
28. **Hamilton, D. B., Walowit, J. A., and Allen, C. M.,** A Theory of Lubrication by Microirregularities, ASME Paper 65-LUB-11, American Society of Mechanical Engineers, New York, N.Y.
29. **Anno, J. N., Walowit, J. A., and Allen, C. M.,** Microasperity Lubrication, ASME Paper 67-WA/LUB-1, American Society of Mechanical Engineers, New York, N.Y.
30. **Sneck, H. J.,** The eccentric face seal with a tangentially varying film thickness, *ASME Trans. Ser. F J. Lubr. Technol.,* 91, 748, 1969.
31. **Sneck, H. J.,** The misaligned, eccentric face seal, *ASME Trans. Ser. F J. Lubr. Technol.,* 91, 695, 1969.
32. **Findlay, J. A.,** Measurements of leakage in mechanical face seals, *ASME Trans. Q., Ser. F J. Lubr. Technol.,* 91, 687, 1969.
33. **Lohou, J. and Godet, M.,** Angular misalignments and squeeze-film effects in radial face seals, in *Proc. 6th Int. Conf. Fluid Sealing,* British Hydromechanics Research Association, Cranfield, Bedford, England, 1973, DD-2-15.
34. **Pape, J. G.,** Fundamental research on a radial face seal, *ASLE Trans.,* 11, 302, 1968.
35. **Stanghan-Batch, B. and Iny, E. H.,** Hydrodynamic theory of radial-face mechanical seals, *J. Mech. Eng. Sci.,* 15, 17, 1973.
36. **Greiner, H. F. and Bussick, F. P.,** Mechanical sealing for heat transfer fluids, *Lubr. Eng.,* 28, 282, 1972.
37. **Lazau, W. R., Shelton, B. R., and Waldheger, R. A.,** The use of carbon-graphite in mechanical seals, *J. Am. Soc. Lubr. Eng.,* 19, 201, 1963.
38. **Trytek, J. J.,** Application of mechanical end face seals for hot water service, *Lubr. Eng.,* 29, 17, 1973.
39. **Pfoutz, B. D., Lagarias, L. S., and Donath, E. A.,** Development of High Temperature Seal-Lubricant Deposit Evaluation Techniques, APL-TDR64-78, September 1964.
40. **Shevchenko, R. P. and Brown, P. E.,** Development of Rubbing Contact Materials for Pratt & Whitney Aircraft Engines, paper presented before the Southern New England Section of the Society of Automotive Engineers, March 14. 1961.
41. **Tankus, H.,** End face seals in abrasive service, *J. Am. Soc. of Lubr. Eng.,* 19, 403, 1963.
42. **Johnson, R. L., Swikert, M. A., and Bailey, J. M.,** Wear of Typical Carbon-Base Sliding Seal Materials for Temperatures to 700°F, NACA Technical Note 3595, 1956.
43. **Campbell, J. G., Dennington, M. G., and Shirodkar, D. N.,** The use of carbon and graphite in slider/counterface systems in the field of mechanical engineering, in *Proc. Industrial Lubrication and Tribology Symp.,* London, November, 1969, Scientific Publication Ltd., 1970, 121.
44. **Brown, P. F., Gordon, N., and King, W. J.,** A test method for evaluating gas turbine engine seal materials, *Lubr. Eng.,* 22, 7, 1966.

45. **Jones, D. A.**, High Temperature Lubricant Screening and Systems Studies, NASA CR-121188, National Aronautics and Space Administration, Washington, D.C., SKF Industries Inc., October 1973.
46. **Hady, W. F., Allen, G. P., Sliney, H. E., and Johnson, R. L.**, Wear and friction of mechanical carbons in liquid oxygen as influenced by transfer films, *ASLE Trans.*, 6, 201, 1963.
47. **Brown, P. F.**, Discussion of Lazau et al. paper, *J. Am. Soc. of Lubr. Eng.*, 19, 207, 1963.
48. **McKibbin, A. H. and Parks, A. J.**, Aircraft Gas Turbine Mainshaft Face Seals. Problems and Promises, Special Publication SP-2, in *ASLE 4th Int. Conf. Fluid Sealing*, Philadelphia, May 5 to 8, 1969, 28.
49. **Dobek, L. J.**, Development of mainshaft seals for advanced air breathing propulsion systems, NASA-CR-121127, March 1973.
50. **Fechter, N. J. and Petrunich, P. S.**, Development of Seal Ring Carbon-Graphite Materials (Tasks 8, 9 and 10), NASA-CR-121092, January 1973.
51. **Mayer, E.**, *Mechanical Seals*, Iliffe Books Ltd., London, 1969, 42.
52. **Snapp, R. B. and Sasdelli, K. R.**, Performance characteristics of a high pressure face seal with radially converging interface shapes in, *Proc. 6th Int. Conf. Fluid Sealing*, British Hydromechanics Research Assoc., Canfield, Bedford, England, 1973, E-4-41.
53. **Hornaday, J. R.**, A History of Small Refrigeration Compressor Shaft Seals, Reprint 65AM3C1 American Society of Lubrication Engineers, Park Ridge, Ill., paper presented at 20th ASLE Annual Meeting, Detroit, Mich., May 4 to 7, 1965.
54. **Schweiger, F. A.**, The performance of jet engine contact seals, *J. Am. Soc. Lubr. Eng.*, 19, 232, 1963.
55. **Teichmann, O. E., Varga, L. A., and Gyory, E. J.**, Experimental techniques in testing liquid sealing face seals, *J. Am. Soc. Lubr. Eng.*, 20, 345, 1964.
56. **Bernd, L. H.**, Survey of the theory of mechanical face seals. I, *Lubr. Eng.*, 24, 479, 1968; Part II, *Lubr. Eng.*, 525, 1968; Part III, *Lubr. Eng.*, 597, 1968.
57. **ASLE Seals Committee**, End Face Seals, American Society of Lubrication Engineers, Park Ridge, Ill., 1968.
58. **Anon.**, Sampling Procedures and Tables for Inspection by Attributes, MIL. Std. 105D, Superintendent of Documents, U.S. Government Printing Office, Washington, D.C., April 29, 1963.
59. **Chiang, T. and Cheng, H. S.**, An analysis of flexible seal ring vibrations, *ASLE Trans.*, 11, 204, 1968.
60. **Hudelson, J. C.**, Dynamic instability of undamped bellows face seals in cryogenic liquid, *ASLE Trans.*, 9, 381, 1966.
61. **Strugala, E. W.**, The nature and causes of carbon seal blistering, *Lubr. Eng.*, 28, 333, 1972.

Chapter 7

CARBON IN BEARINGS

I. USES OF CARBON BEARINGS

In an estimated 3% of all bearing applications, performance could be improved and costs reduced by the use of carbon. The purpose of this chapter is first to point out situations where carbon bearings should be used. Following that, methods of design and materials selection will be reviewed.

Listed below are 14 situations where carbon bearings should be used:

1. Temperature above 200°C (400°F) cokes oil and grease
2. Temperature below −40°C (−40°F) congeals oil
3. Pressures below 10^{-3} N/m^2 (10^{-8} atm) vaporize oil
4. Radiation flux decomposes oil
5. Centrifugal or vibrational forces throw off oil
6. Surrounding solvents wash away oil
7. Corrosive fluids attack bearing metals
8. Periodic sterilization (or steam cleaning) removes oil
9. Oil would contaminate the surrounding product
10. Dusty environment contaminates oil
11. Vibration or oscillation of bearing brinels bearing races
12. High static loads, with occasional movement, squeezes out oil
13. Bearing location inaccessible for periodic lubrication
14. Fail-safe operation required

A. High Temperatures

Carbon should be considered if the bearing will have to operate above 200°C (400°F) for more than a few hours during its life. At this temperature most oils and greases start to decompose in the presence of air. Plastic materials such as PTFE (Teflon®, etc.) soften and lose their load-carrying capacity. Carbon, in contrast, actually grows stronger and harder with increasing temperature; it doubles its room temperature strength when it is heated to 2200°C (4000°F).

Many successful applications confirm the effectiveness of carbon at high temperature. Kellogg and Dewart[1] present test results on carbon bearings used at temperatures to 620°C (1150°F) on the control drums of the SNAP-8 reactor. Carson[2] pictures a combination thrust and radial carbon-lined bearing used to support variable stator blades in the compressor section of an aircraft jet engine. Performance data on this type of carbon bearing operating at temperatures as high as 650°C (1200°F) are reported by Paxton and Scilingo.[3] Many industrial installations use carbon bearings on conveyors, in kilns, and on fans operating at temperatures between 200 and 675°C (400 and 1250°F).

In general, the coefficient of friction and the wear rate of carbon rubbing against metal increases with increasing temperature. Johnson et al.[4] found the increase in wear rate to be small below 260°C (500°F) but observed that the rate climbed sharply at higher temperatures. Rabinowicz,[5] Rabinowicz and Imai,[6] Sibley et al.,[7] Swikert and Johnson,[8] Lotos,[9] and Bisson et al.[10] extended this investigation to 1095°C (2000°F) and observed similar trends. Driesner and Wagner[11,12] made friction measurements of graphite rubbing against graphite at temperatures up to 2450°C (4450°F) and under loads up to 17 MN/m^2 (2460 lb/in.^{2}). Even at these extreme conditions the coefficient of friction was below 0.5, and rubbing took place without galling or seizing.

At temperatures above 325°C (620°F), ordinary carbons react slowly with air and eventually lose their strength due to oxidation. Thus, carbons intended for long-term exposure at higher temperatures generally contain oxidation inhibitors, commonly phosphorous[13-15] or boron[16] compounds. Many of these compounds also act to lower friction and wear rates. Paxton and Shobert,[17] Shobert and Olson,[18] and Swikert and Johnson[8] report friction coefficients dropping well below 0.1 and find the wear rates reduced to 1/100 the "normal" value at temperatures above 425°C (800°F). Carson[2] lists the commercial designations and suppliers of a number of these materials. These oxidation inhibitors, when applied to the proper carbon grades, are very effective, giving thousands of hours of useful life to carbon bearings operating in air at 675°C (1250°F).

B. Low Temperatures

Below −40°C (−40°F) most oils congeal or become so viscous that bearing power losses rise sharply. Carbon bearings utilize the cryogenic fluid itself as a lubricant and coolant. Being self-lubricating, carbon bearings can operate when they are completely dry without damaging the rotating shaft.

The potential of carbon as a rubbing material in cryogenic systems was first explored in depth by Wisander et al.[19] He discovered that at cryogenic temperatures certain impregnated grades of carbon had wear rates only 1/100 those of other carbon grades. His work was extended by Bisson[20] and Hady et al.[21,22] who reported the best performance was obtained with impregnated carbons that have a high graphite level. When high-graphite carbons were rubbed against hard chrome plate on the mating metal surfaces, minimum wear rates were produced. As a result of this pioneering work, carbon bearings and seals are now widely used in liquid oxygen and liquid hydrogen pumps.

C. Low Pressures (Hard Vacuums)

Bearings for devices that are to be used outside the atmosphere of the earth or in high-vacuum chambers present special problems. The greases or lubricants normally used evaporate under these conditions, and metal to metal contact generally results in adhesion. Because of the special properties of carbon, bearings made from it are used in many space applications. The SNAP-8 bearings,[1] which operate in a 10^{-3} to 10^{-8} N/m^2 (10^{-5} to 10^{-10} torr) environment, are but one example.

Carbon depends on adsorbed gases for its lubricating ability, a property which at first glance would apparently rule out its use in vacuum applications. Buckley and Johnson[23] observed wear rates increased 1000-fold when the pressure was dropped from 1 atmosphere to 10^{-5} N/m^2 (10^{-7} torr). The use of impregnants such as silver or BaF_2 so improved the performance of the carbon, however, that wear rates at 10^{-7} N/m^2 (10^{-9} torr) were only slightly greater than those at 1 atmosphere. Kellogg[24] screened 67 material pairs at 10^{-7} N/m^2 (10^{-9} torr), 700°C (1300°F) and a load of 0.28 MN/m^2 (40 lb/in.^{2}) and concluded that impregnated carbon was the only workable bearing material.

D. High Radiation Fluxes

High radiation fluxes adversely affect most greases, oils, and plastics. Carbon, on the other hand, is virtually unaffected by α (proton), β (electron) or γ (X-ray) radiation. Neutron radiation is also without measurable effect until exposure dosages exceed 10^{20} neutrons per cubic centimeter. The SNAP-8 carbon bearings,[1] for example, are exposed to 10^{11} rad γ and 10^{20} fast neutrons per cubic centimeter during the 10,000-hr working life. At thermal neutron dosages above 10^{21} neutrons per cubic centimeter, cumulative damage to the crystal lattice of carbon can cause shrinkage up to 10%. Fast neutrons have far less effect on the lattice.

FIGURE 7.1. Metal shrouded bearing for use at 425°C (800°F) and above.

E. High Centrifugal Forces

Helicopter rotor control bearings are typical examples of bearings that must operate in high centrifugal force fields. Oil- or grease-lubricated bearings quickly lose their lubricant. Plastic-lined bearings yield because of repeated load reversals and give excessive clearance in the bearing. Carbon, however, needs no lubricant and does not yield or cold-flow. Advanced helicopters now being developed will use many carbon-lined bearings.

F. Solvents, Water, or Corrosive Fluids

There are many applications where bearings must operate submerged in gasoline, jet fuel, or other organic liquids. These liquids can dissolve and leach out oil or grease lubricants from rolling element or metallic sleeve bearings. Examples of carbon use for bearings in aircraft fuel pumps are embodied in the patents of Nickerson.[25,26] A recent development is the use of carbon bearings in Freon compressors. Premature failures of the main shaft bearing occurred on about 5% of the compressors shipped by a major producer. The problem was traced to loss of oil from the sintered iron bearing. This loss was caused by the "vapor degreasing" action of condensing freon while the compressors were idle. Galling occurred when the compressor first started and continued until oil laden liquid freon again saturated the bearing. By then the damage was done. Simple substitution of a carbon sleeve for the sintered iron sleeve solved this problem.

The use of water- or steam-lubricated carbon bearings on pumps and turbines provides another obvious application. Vitellozi[27] explored a number of shaft-bearing combinations and obtained the best results with a hardened 17.5% chromium-1% carbon shaft running against a silver-impregnated carbon bearing. He explored both steam and warm water (46°C = 115°F) as lubricants, loads from 0.14 to 0.3 MN/m^2 (20 to 45 $lb/in.^2$) and rotational speeds from 1,000 to 15,000 rpm on a 38 mm (1 1/2 in.) bore bearing.

Cornell et al.,[28] after an extensive experimental investigation, concluded that hard carbon running against a chrome-plated shaft offered the best combination of rubbing members for use on relatively highly loaded, slowly rotating ship propellor shafts. Summers[29] described the design of carbon thrust bearings for large 670 kW (900 hp) submerged motors.

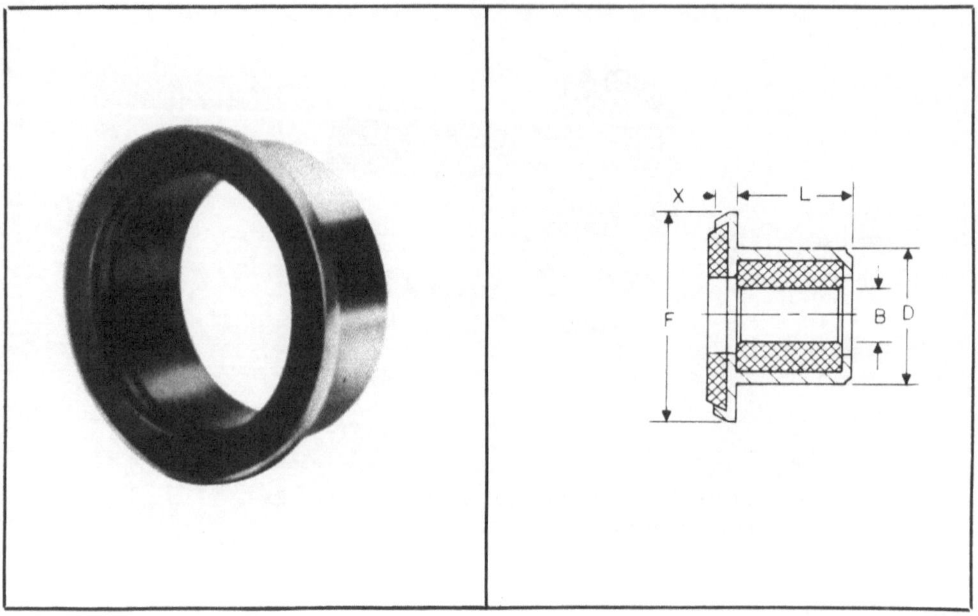

FIGURE 7.2. Photo and sketch of combination radial and thrust bearing that can operate submerged in gasoline, water, or corrosive fluids.

In still other cases, bearings must operate submerged in corrosive fluids such as hydrochloric or hydrofluoric acid. Here, too, carbon bearings offer a commonly used alternative to sealed and pressurized rolling element bearings.

G. Repeated Steam Cleaning or Sterilization

Food-handling equipment and medical devices are examples of apparatus that must be repeatedly steam cleaned or sterilized. Carbon bearings, particularly those made from impervious, nonporous carbon grades, offer significant advantages over grease- or oil-lubricated bearings. They are unaffected by the high temperatures or gases used for sterilization, and they are immediately ready to reuse without regreasing.

H. Applications Where Product Contamination Must Be Avoided

There are many uses for carbon bearings in food-handling equipment — such as drink dispensers, mixers and blenders, sanitary hardwear, and food-wrapping devices — where oil contamination must be avoided. Oil contamination cannot be tolerated on fabrics either; hence, carbon bearings are used in looms and weaving equipment. Since carbon is unaffected by body fluids, carbon bearings are used in blood pumps, artificial kidneys, and even prosthetic devices implanted in the human body.

I. Bearings for Dusty Environments

Keeping dust, lint, or other solids out of greased bearings is a common problem. In some cases — such as on farm machinery, rock conveyors, or textile machines — it is cheaper to use dry, self-lubricating carbon bearings and omit the dust seals, the grease fittings, and the cost of periodic greasing.

J. Vibrating or Oscillating Bearings

Carbon does not yield, cold-flow, or work-harden under load; hence, it can be expected to carry repeated loads without damage. Moreover, the low modulus of carbon relative to that of most shaft materials protects the shaft from localized high-Hertzian contact stresses by distributing the loads over a wide area. Accordingly, carbon-lined journals are ideal for those applications where vibration or oscillation can lead to brinelling the races of ball or roller bearings. A typical application that takes advantage of these attributes is control linkage bearings on fixed wing aircraft. These are often subjected to pulsating loads caused by aerodynamic flutter. Figure 7.3 shows a typical aircraft bearing. Some of these bearings are used to carry projected loads in excess of 138 MN/m^2 (20,000 lb/in.^{2}).

K. High-Load Bearings

Occasionally bearings are required to carry high static loads for long periods of time followed by minor slow movement. To carry the high static load, massive and costly rolling element bearings may be used. If, instead, a less expensive, greased metal to metal bearing is used, there is always the danger that the grease will be squeezed out and the bearing will seize when movement is called for. Carbon bearings, which can carry static loads in excess of 172 MN/m^2 (25,000 lb/in.^{2}) of projected area with no added lubricant, are ideal for these applications.

L. Inaccessible Bearings

The lubrication of bearings in inaccessible locations is often neglected; hence, self-lubricating carbon bearings are preferable. Typical applications are found on military devices and farm equipment where good maintenance is difficult to obtain.

M. Fail-Safe Applications

Carbon will not gall or freeze; hence, it is used for bearings where failure cannot be tolerated. Examples that require fail-safe bearings can be found in the hydraulic pumps and drives used in most large aircraft.

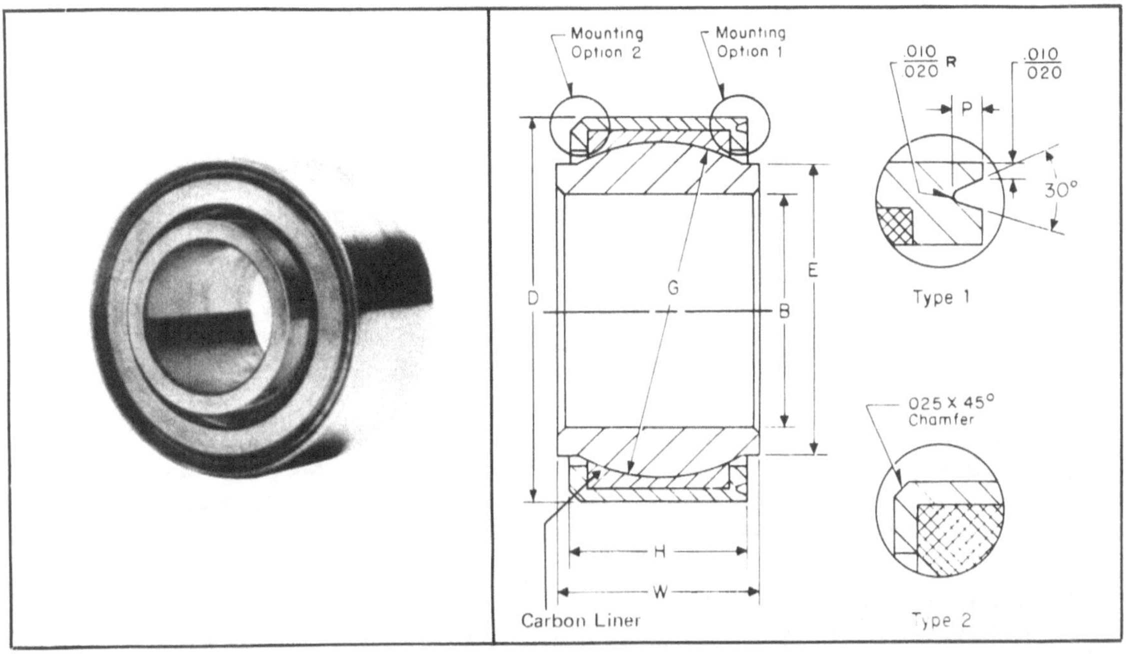

FIGURE 7.3. A photo of aircraft control linkage bearings. (All dimensions in inches.)

II. STATIC LOAD CAPACITY OF BEARINGS

Where the lifetime linear movement of shaft surface across a given point on the carbon bearing surface is small, say less than 25 m (1000 in.), carbon wear can be ignored, and the bearing can be designed solely on the basis of static capacity. Static capacity, in turn, depends on shaft stiffness, carbon compressive strength, bearing design, and the definition of failure.

For our purposes, failure will be defined as the appearance of cracks in the carbon liner, chips around the circumferences at either end of the carbon liner, or a permanent set in excess of 0.3% of the bearing bore.

A. Static Capacity — Sleeve Bearing

Figure 7.4 shows typical sleeve bearing configurations, one with and one without axial retainer plates. Figure 7.5, taken from Paxton and Scilingo,[3] shows the testing device used to find the effect of the compressive strength of carbon and the carbon liner thickness upon static capacity of the bearing assembly. Figure 7.6 sketches the appearance of the load-bearing half of the liner at various stages of failure. The shaft used in these tests was hardened tool steel, modulus about 275 GN/m^2 (40 × 10^6 lb/in.^{2}), yet it is obvious that shaft deflection was giving localized overloading and producing failure due to chipping at the bearing ends.

Figure 7.7 sketches a simplified force diagram for the test fixture of Figure 7.5. Calculated shaft bending is 2.0 × 10^{-3} mm (78 × 10^{-6} in.), enough to add 75 MN/m^2 (10,400 lb/in.^{2}) or 15% to the average projected load on the edges of the 0.76-mm thick carbon liner. Actual bending moment on the shaft would be greater than that calculated in Figure 7.7 because some of the force is distributed over a longer moment arm than shown in the simplified diagram. It follows that actual shaft deflection and concentrated excess end loads are even higher than the calculated value.

Figure 7.8 summarizes the findings of Paxton and Scilingo.[3] Static radial load capacity was always at least as high as that calculated using carbon compressive strength and projected bearing area. Increasing carbon liner radial thickness up to 2.5 mm (0.100 in.) increased load-carrying capacity. Capacity improved as much as 50% with high-modulus carbons, less with the more compliant, low-modulus grades.

B. End Plates for Radial Sleeve Bearings

When thick carbon liners are used, there is always the danger that if the carbon cracks it will fall out and totally destroy the bearing. This possibility can be virtually eliminated by applying metal retainer plates to the ends of the bearing. Figure 7.4 shows a commercially available carbon bearing which uses this approach. Figure 7.9 shows the relation between load and total deflection for this assembly. End chipping did not appear until the load (projected) reached 525 MN/m^2 (76,000 lb/in.^{2}). This corresponds to 165% of the carbon compressive strength determined by conventional (ASTM C-695) methods.

Another 15 bearings using the design of Figure 7.4 and ranging in bore up to 25.4 mm (1.00 in.) were statically loaded until end chipping appeared. Six grades of carbon were used as liners in this series of tests. Chipping appeared at projected loads ranging from 100 to 260% of the compressive strength of the liner material.

C. Static Capacity — Thrust Bearings

If sleeve bearings can carry static loads, based on projected areas, ranging from 100 to 260% of the compressive strength of the carbon, one would expect thrust bearings to show even higher multiples. This does not happen in present designs.

Figure 7.10 shows a typical thrust bearing design. The carbon is held in place by a

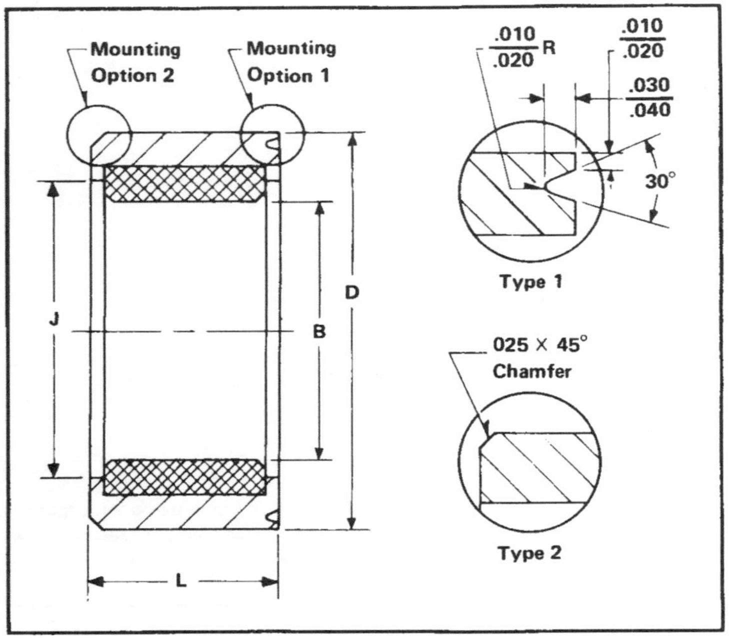

FIGURE A.

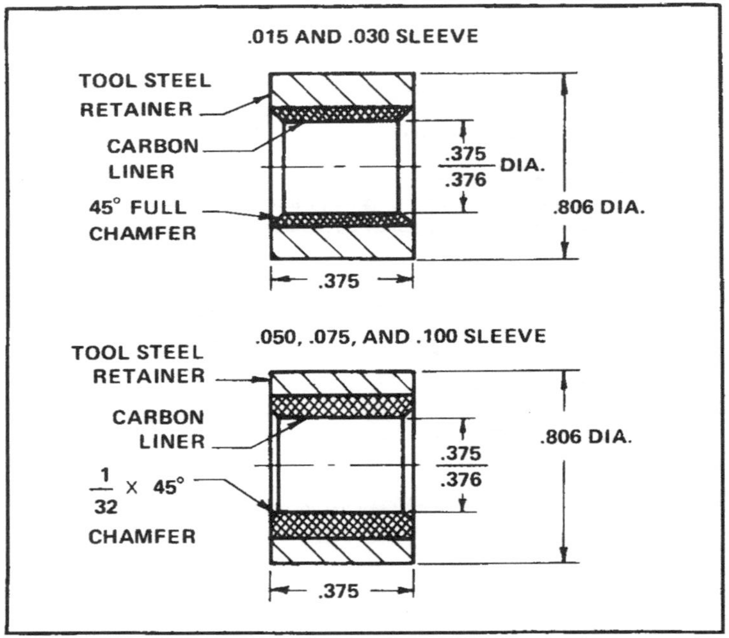

FIGURE B.

FIGURE 7.4. Sleeve bearings. (A) With side plates. (B) Plain sleeve bearing. (All dimensions in inches.)

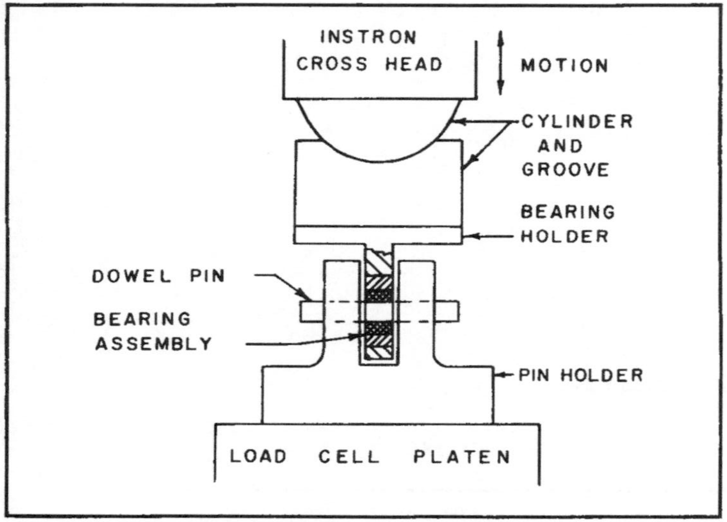

FIGURE 7.5. Test fixture for determining static load capacity of a carbon-lined sleeve bearing. (From Paxton, R. R. and Scilingo, A. A., *Lubr. Eng.*, 25, 246, 1969. With permission.)

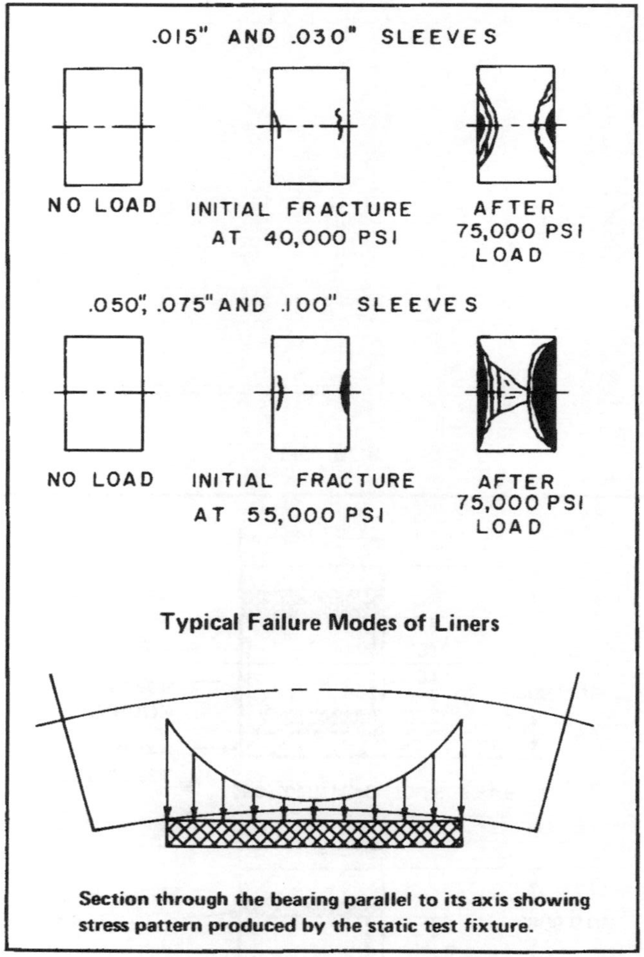

FIGURE 7.6. Failure mode during static tests of sleeve bearings having carbon liners of various thickness. Appearance of the inner surface of the loaded half of the bearing. (Dimensions in inches.)

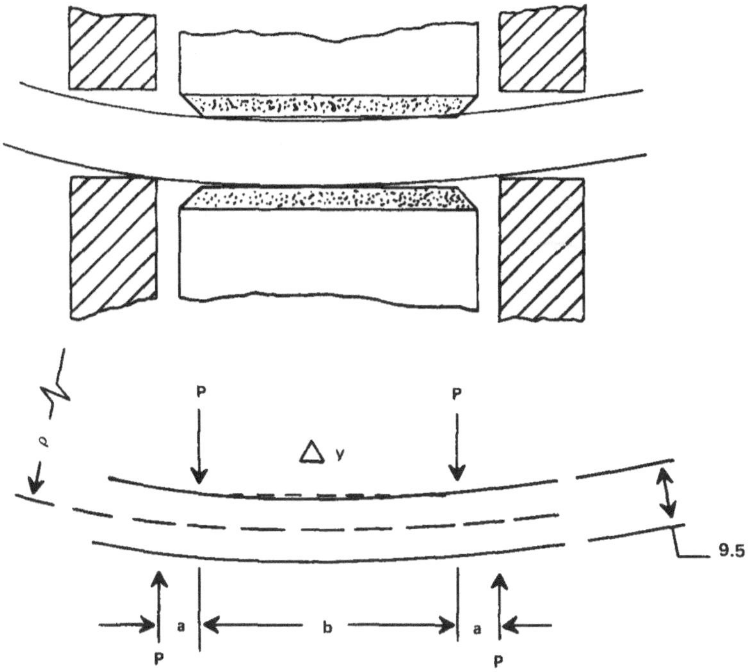

a = 1.52 mm
b = 8.0 mm
P = 17.79 kN

E rod = 275 GN/m^2 (40 x 10^6 lb/in^2)
E carbon liner = 27.5 GN/m^2 (4 x 10^6 lb/in^2)
L, carbon liner thickness = 0.76 mm (0.020 in.)

Average Projected Load = 2P/9.5 b = 2(17.8)/9.5 (8.0) = 470 MN /m^2 — Eq. 7.1

Bending Movement, M = Pa = 17.79 (1.52) = 27 Nm — Eq. 7.2

Max Fiber Stress in Rod, s = M/Z

Max. Fiber Stress in Rod, $s = M/Z = \frac{32M}{\pi d^3} = \frac{32(27)}{\pi(9.5)^3} = 320$ MN/m^2 (46,700 lbs/in^2) — Eq. 7.3

where $Z = \frac{\pi d^3}{32}$ — Eq. 7.4

Radius of curvature of rod, $\rho = \frac{E}{s} \cdot \frac{d}{2} = \frac{(275{,}000)\ (9.5)}{(320)\ (2)} = 4.08$ m — Eq. 7.5

Rod deflection, $\Delta y \cong \frac{(b/2)^2}{2\rho} \cong \frac{(8.0/2)^2}{2 \times 4{,}080} = 0.002$ mm — Eq. 7.6

Localized *added* load due to this deflection, $\Delta = (E\ \text{carbon})\ \frac{\Delta Y}{L} = \frac{(27.5 \times 10^9) \times (2 \times 10^{-6})}{0.76 \times 10^{-3}}$

$= 72$ MN/m^2 (10,400 lb./in.^{2}) — Eq. 7.7

FIGURE 7.7. Calculation of bearing end load caused by shaft bending.

rolled metal belt. The bearing is tested by loading the carbon face. As in the case of sleeve bearing tests, "failure" was defined as the appearance of chips or cracks in the carbon. In most cases, however, the bearing remained fully functional even though it had "failed".

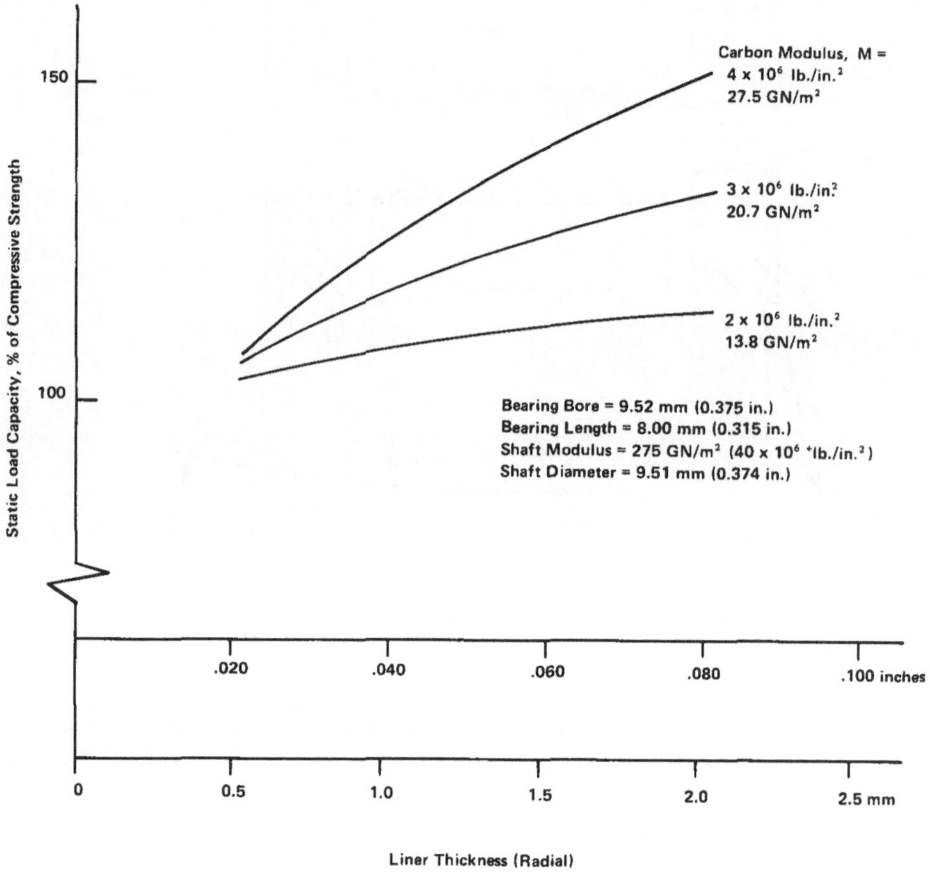

FIGURE 7.8. Static load capacity increases as carbon liner thickness increases.

Figure 7.11 is a load-deflection curve for the bearing shown in Figure 7.10. Figure 7.11 shows the test fixture used to obtain the data for Figure 7.10. Calculated static capacity is the product of the contact area and the compressive strength of the carbon as determined by ASTM C-695. Actual static load to "failure" as defined above ranged from 103 to 200% of the calculated value for five test bearings for each of four grades of carbon.

D. Uniball Bearings — Radial Static Capacity

Figure 7.12 shows a typical "uniball" bearing assembly. Two carbon rings having spherically ground inner surfaces are assembled in a metal retainer. A spherical metal ball is trapped in place by the spherically ground carbon members. The ball, of course, may be a hardened tool steel such as M-2, or it may be hard-surfaced with chrome plate or flame-sprayed ceramics.

The thickness and stiffness of the metal shell confining the carbon elements may vary as may the axial gap between the two carbon elements. Another variable is the grade of carbon used. Carbons, of course, vary in compressive strength as well as coefficient of friction between the ball and the carbon supports.

When a bearing such as that shown in Figure 7.12 is loaded radially, there is a tendency for the carbons to move apart axially. This, in turn, increases the clearance between the ball and its support. Failure, therefore, may arise either from excessive clearance or from crushing the carbon supports.

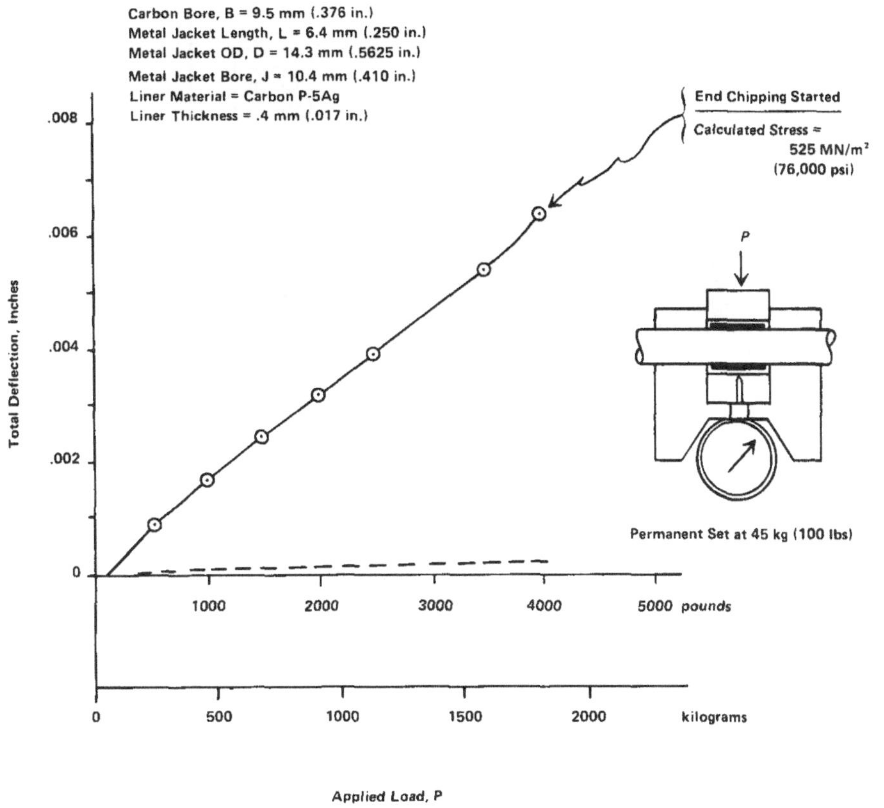

FIGURE 7.9. Load-deflection curve for carbon bearing.

Eleven bearings with ball diameters ranging from 15.9 (0.625 in.) to 26.2 mm (1.030 in.) were made up according to Figure 7.12. Axial thickness of the carbon plus its metal container ranged from 8.5 mm (0.335 in.) to 12.7 mm (0.500 in.). The thickness of the steel used to house the carbon parts was 5% of the ball diameter, G. Each bearing was placed in the fixture shown in Figure 7.13 and loaded in gradually increasing steps until "failure" occurred. Failure was defined as an increase in radial clearance of 0.15% of the ball diameter or chipping of the carbon support, whichever occurred first. Starting clearance and final clearance were measured at a preload of 0.25% of the calculated radial load capacity. Expected radial load capacity was calculated from the formula:

$$RLC = 0.5\ GHS \tag{7.8}$$

RLC = 0.5 GHS where RLC = radial load capacity in Newtons force; G = ball diameter in meters; H = axial thickness of the support cartridge in meters; and S = compressive strength of the carbon liner in MN/m² per ASTM C-695.

Figure 7.14 shows the results of this test series. Ten of the 11 bearings "failed" because clearance exceeded the allowable increase of 0.15% of ball diameter.

A second series of bearings having thinner carbon liners and heavier metal retainers was also tested. In this series, the steel used to house the carbon had a thickness equal to 10% of the ball diameter. Ball diameters on these bearings ranged from 15.1 to 37.2 mm (0.593 to 1.500 in.), and cartridge width varied from 10.2 to 20.2 mm (0.401 to 0.797 in.). Failure criteria were the same as in the above case.

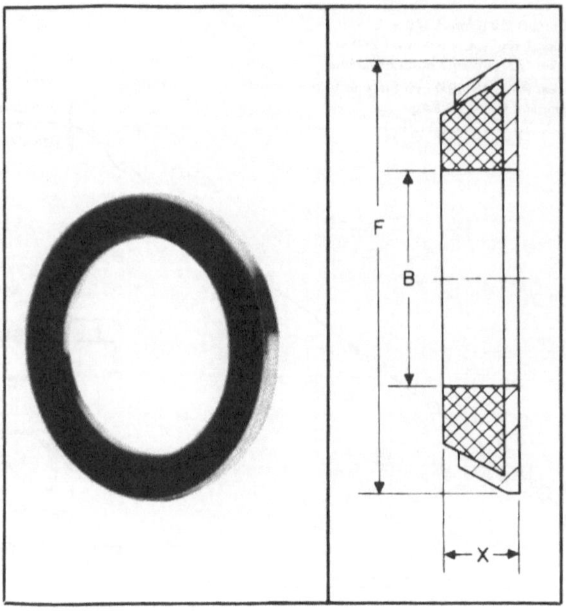

FIGURE 7.10. Metal jacketed thrust bearing. Test bearing size: B = 9.19mm (0.386 in.) F = 22.2mm (0.875 in.) X = 1.59mm (0.0625 in.).

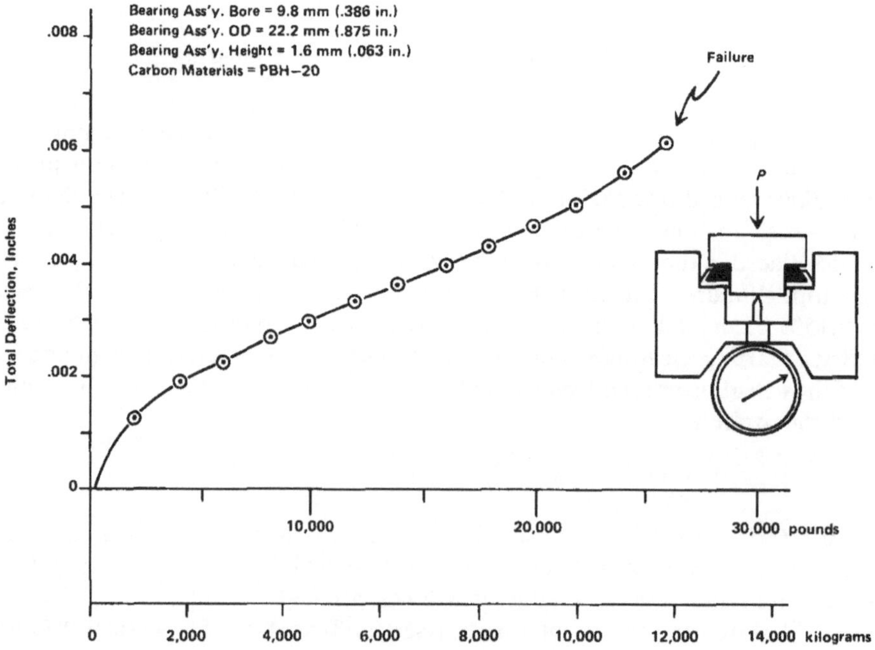

FIGURE 7.11. Load-deflection curve for carbon thrust bearing.

Figure 7.15 compares the load at failure to that calculated from Equation 7.8. With the heavier metal design, only seven of the 11 "failed" because clearance increase exceeded the allowable 0.15%. When the heavier design is used, the probability is better than 0.99 that the radial static load capacity will exceed that calculated by using Equation 7.8.

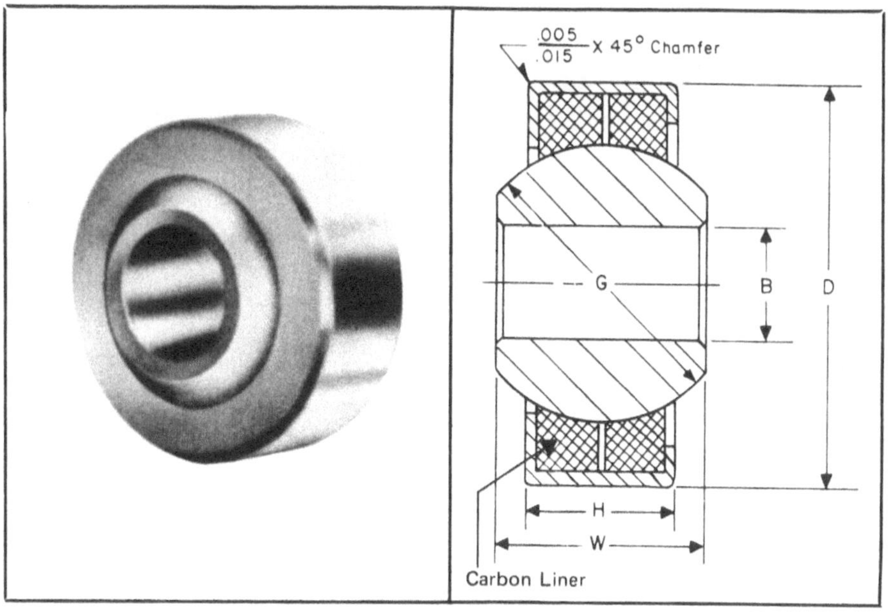

FIGURE 7.12. Light duty spherical bearing. (All dimensions in inches.)

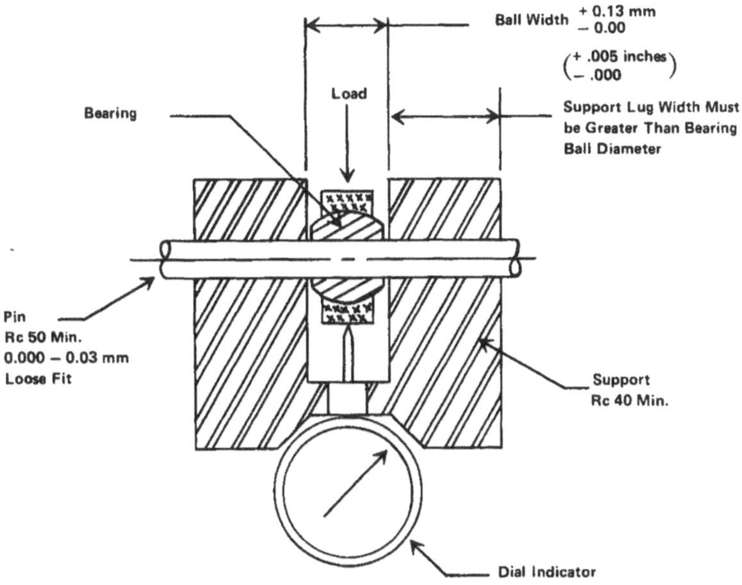

FIGURE 7.13. Fixture for determining radial static load capacity of spherical bearing assemblies.

A comparison of Figures 7.14 and 7.15 shows that the average radial load capacity is increased about 60% by using a more rigid metal retainer construction. Further increases in rigidity of the retainer beyond that of Figure 7.15 would probably show little benefit since carbon cracking rather than excessive clearance is beginning to cause "failure."

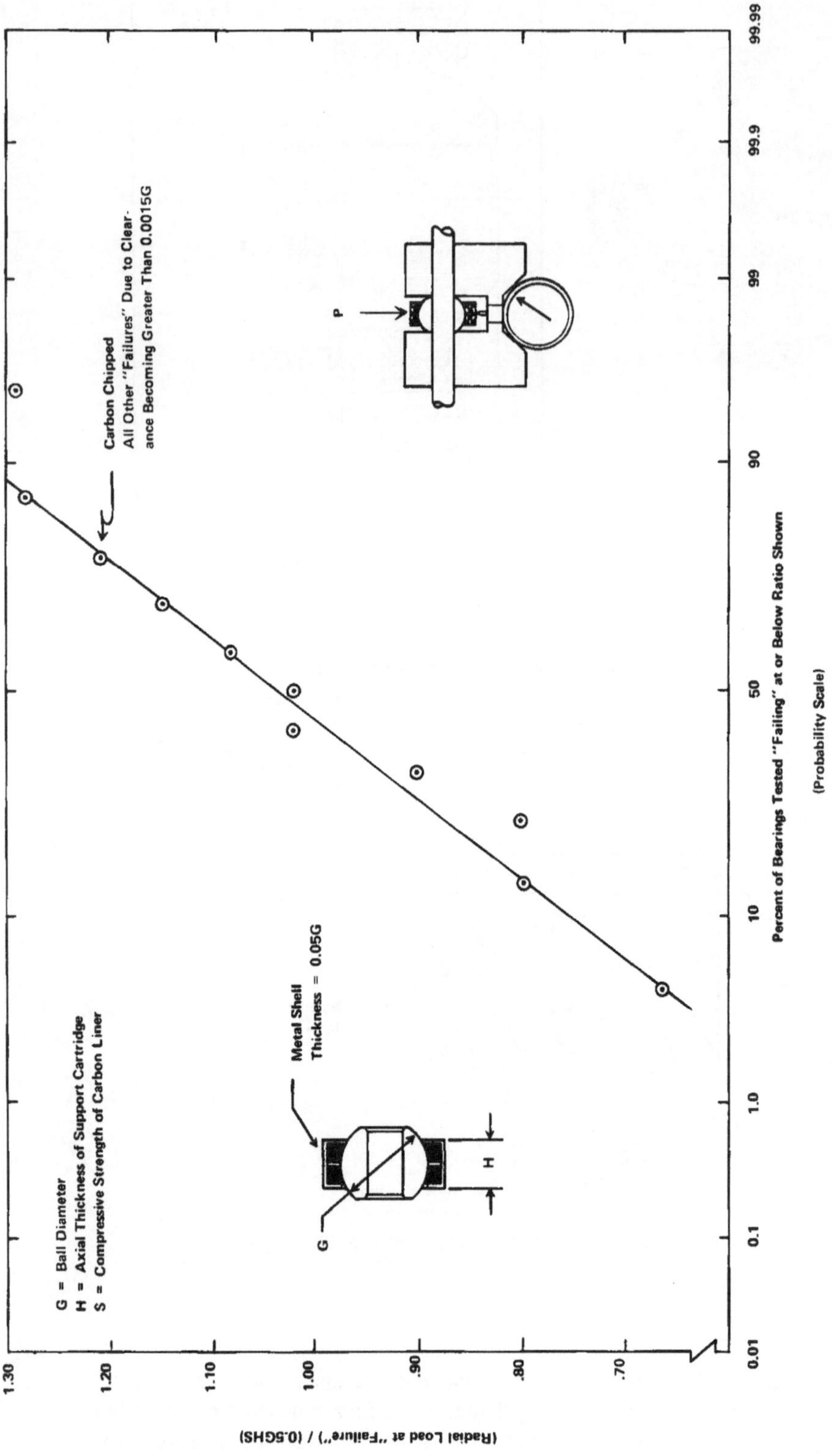

FIGURE 7.14. Radial static load capacity of light duty spherical bearings.

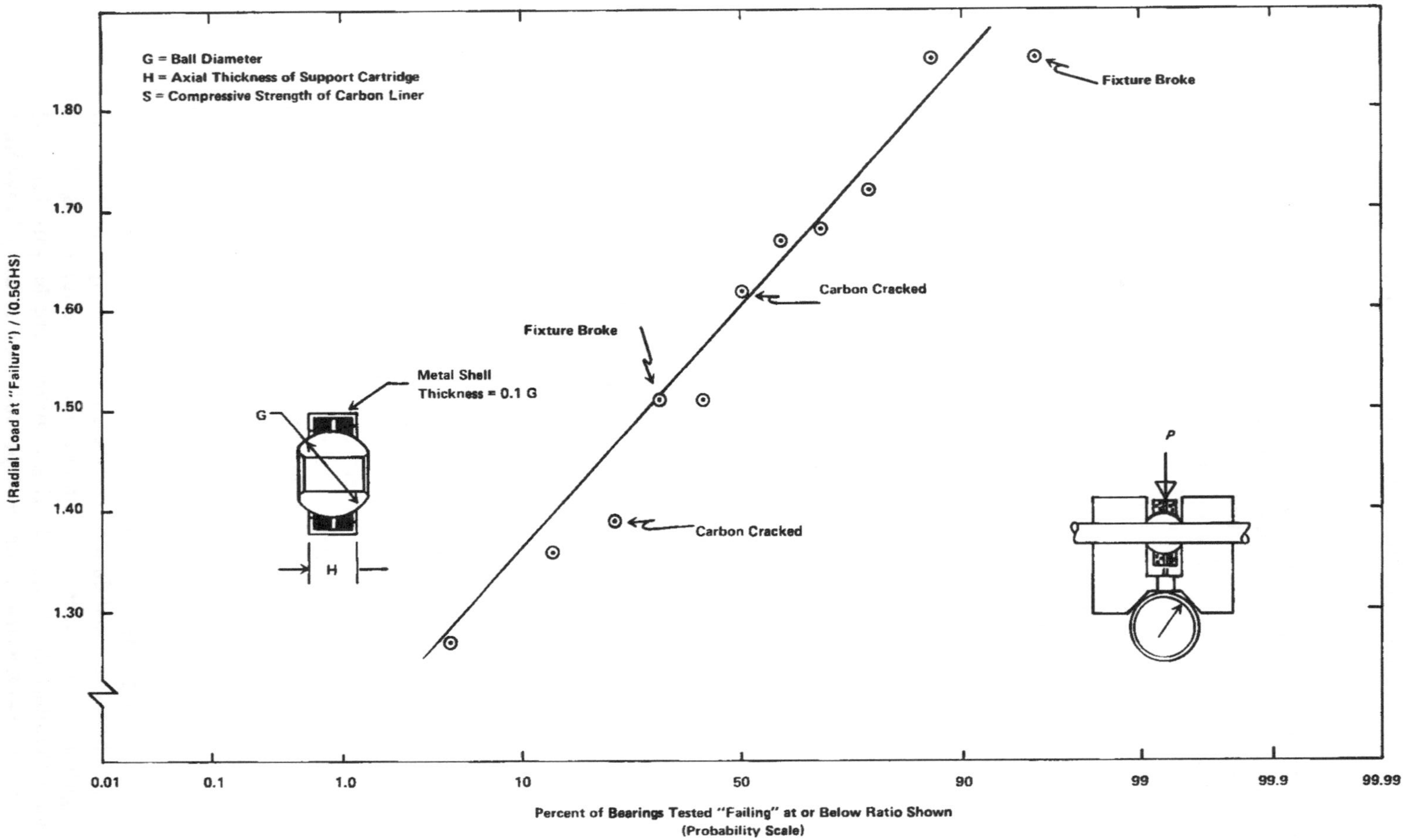

FIGURE 7.15. Radial static load capacity of heavy duty spherical bearings.

A comparison of Figures 7.14 and 7.15 shows that the average radial load capacity is increased about 60% by using a more rigid metal retainer construction. Further increases in rigidity of the retainer beyond that of Figure 7.15 would probably show little benefit since carbon cracking rather than excessive clearance is beginning to cause "failure."

E. Uniball Bearings — Axial Static Capacity

Axial loading of uniball bearings tends to shear the spherical carbon socket. In addition, there is a radial vector whose magnitude depends on the axial load, the coefficient of friction between the carbon socket and the ball, and upon the retainer width, H, relative to ball diameter, G. This radial vector places a tensile hoop stress on the carbon socket. The carbon socket is restrained from radial expansion by the tightly fitted circumferential metal retainer. Thus, the hoop stress generated by axial loading is ultimately accepted by the metal retainer, and the forces in the carbon socket become outwardly compressive in the radial direction and axially shear.

Shear strength of mechanical-grade carbons can be taken as approximately 40% of the compressive strength. Using the form of Equation 7.8, one can estimate the axial load capacity as:

$$ALC = 0.2\ GHS \qquad (7.9)$$

where ALC = axial load capacity in Newtons force; G = ball diameter in meters; H = axial thickness of support cartridge in meters; and S = compressive strength of carbon liner in MN/m^2 per ASTM C-695.

To test this formula, 16 spherical surface bearings were made and tested. Nine of the bearings had lightweight metal shells, and seven bearings had the heavier shell construction. The bearings were then placed in a fixture which supported the shell. The ball edge extending from the opposite face was loaded to a force of 0.25% that of the calculated capacity (Equation 7.9). With this load applied, the deflection gage was set at zero.

Above 50% of the calculated capacity, the loading was increased in increments of 10%. After each increment, the bearing load was reduced to 0.25% of calculated capacity, and the permanent deformation was read. The bearing was then removed from the test fixture, and the carbon liner was inspected for chips or cracks. If none were found, it was returned to the test fixture, and the next increment of load was applied. Failure was defined as the appearance of chips or cracks in the carbon or as a permanent axial deformation of 0.4% of the ball diameter.

Figure 7.16 shows the results of these axial load capacity tests. When the heavy metal retainer was used, the median failure was 50% above the calculated capacity, and four of the seven failures were due to chipping or cracking the carbon liner. When the light metal retainer was used, all failures were due to permanent deformation.

It is important to note that the light metal retainer is not symmetrical in the axial direction (see Figure 7.12). One face of the housing gives more support to the carbon liner than the other. As a consequence, the load at "failure" depends on which way the bearing is placed in the support fixture. In this test program, loading direction was chosen at random. Load at "failure" averaged 75 and 125% of average load capacity (ALC) in the weaker and stronger direction, respectively.

F. Static Load Capacity — Summary

The preceding pages have presented data on the static load-carrying capacity of carbon bearings. In general, these values are far higher than those for rolling element bearings of similar dimensions and weight. Bearings, however, are designed for movement as well as carrying loads. Carbon can be depended upon to carry heavy loads and to permit sliding movement without galling or welding. But it does wear and it does generate frictional heat. The next section will be a discussion of the relation between application conditions and the life of carbon bearings.

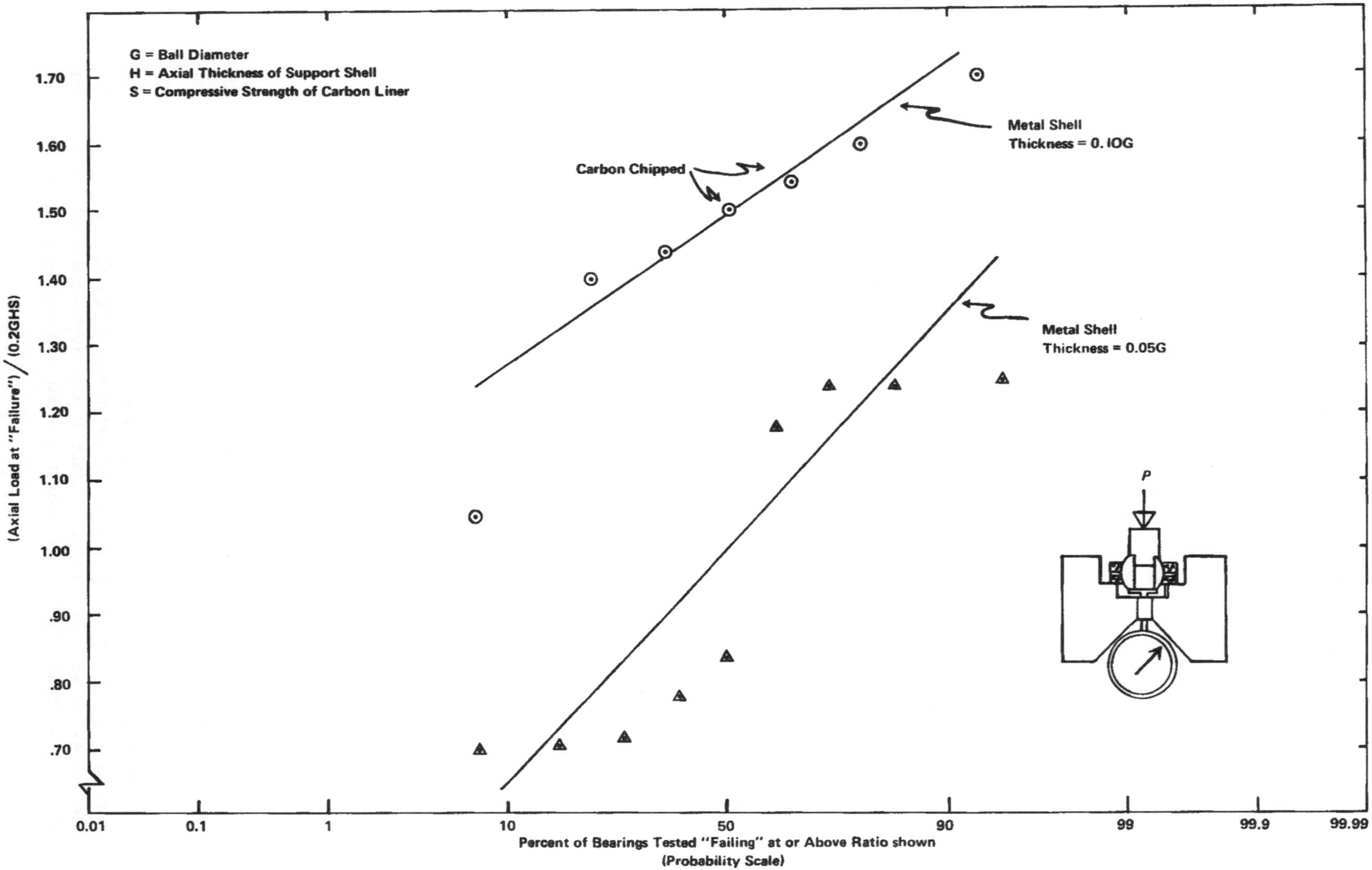

FIGURE 7.16. Axial static load capacity of spherical bearings.

III. USEFUL LIFE OF CARBON BEARINGS

Some factors which affect the useful life of carbon bearings are listed below:

1. Allowable wear
2. Load
3. Rubbing velocity
4. Shaft material, hardness, and finish
5. Temperature, both ambient and bearing surface
6. Fluid environment
7. Alignment and runout
8. Bearing size
9. Carbon grade used
10. Unidirectional or reversing movement
11. Unidirectional or reversing loads
12. Vibration amplitude and frequency

Data on the wear rate and its relation to even these listed factors are virtually nonexistent. Kellogg and Dewart[1] report "satisfactory performance" of a carbon journal after 5000 hr of testing at 620°C (1150°F) in a 10^{-5}-torr vacuum but give no wear values. Paxton and Shobert,[17] Shobert and Olsen,[18] and Carson[2] report wear rates for carbon thrust bearings, but there is no accepted correlation between these data and wear rates for sleeve bearings, for example.

Tests conducted in my laboratory on radial and spherical journals show a total wear of 0.2 mm (0.008 in.) or less after 7500 m (25,000 ft) of rubbing, provided the following limits are observed:

Shaft hardness	Above 35 R_c (Rockwell C)
Shaft finish	Below 0.7 μm AA (20 μin. AA)
Rubbing velocity	Below 0.025 m/sec (5 ft/min)
Load	Below 25% of calculated radial capacity
Environment	Air, 0 to 120°C (0 to 250°F)
Carbon grade	Most carbon-graphites satisfactory

Operation over a far wider range of conditions with wear rates an order of magnitude lower can be achieved by selecting the proper carbon grade.

A. Allowable Wear

Allowable wear is usually set by performance requirements of the design. Low speed (below 100 rpm) journals can generally tolerate radial wear up to 1% of the journal diameter or 0.2 mm (0.008 in.), whichever is greater. Spherical bearings subjected to reversing loads can tolerate only one third of this wear. High-speed (above 1000 rpm) journal bearings can usually tolerate radial wear up to 0.5% of journal diameter or 0.1 mm (0.004 in.), whichever is greater.

Allowable wear for thrust bearings depends on permissible axial movement of the support member. Wear limits as high as 0.6 mm (0.025 in.) are not unusual.

B. Load

Bearings are often rated in terms of allowable PV levels in which "P" is projected load, and "V" is the rubbing velocity. This rating system postulates a limiting permissible energy release at the rubbing interface. It implicitly assumes that the coefficient of friction at the interface is independent of pressure and rubbing velocity. With carbon, neither portion of this assumption is true. The friction coefficient of carbon generally decreases with increasing load and with increasing speed.

In my laboratory, wear rates of carbon-lined sleeve bearings and of spherical bearings have been measured at loads ranging from a few kiloNewtons per square meter (pounds per square inch) to 70 MN/m^2 (10,000 $lb/in.^2$). Even at the highest loads, wear values as low as 0.005 metres of wear per 1 million metres rubbed were observed when the proper carbon grade and journal surface were mated. Data from a typical example with a moderate load are shown below:

Carbon grade	P-03XHT
Journal material	17-4 PH stainless steel
Journal hardness	R_c = 44
Journal finish	0.3 to 0.4 μm AA (12 to 16 μin.)
Rubbing speed	0.36 m/sec (70 ft/min)
Contact pressure	2.6 MN/m^2 (375 $lb/in.^2$)
Ambient fluid	Hot air
Bearing temperature	300°C (572°F), by measurement
Wear ratio	0.8×10^{-9}
Wear rate	0.001 mm/hr (0.00004 in./hr)
Friction coefficient	0.05 to 0.1

C. Velocity

Many investigators fail to recognize the importance of temperature rise when obtaining wear rate data at various rubbing velocities and loads. Wear rate is strongly influenced by the temperature of the rubbing surfaces, and the temperature of the rubbing surfaces is established by a dynamic balance between heat generation and heat loss. Heat loss rates depend on geometry, materials, and fluid environment near the rubbing surfaces. Heat generation depends upon rubbing speed, load, and coefficient of friction.

When obtaining wear data, care should be taken to separate temperature effects from those caused by load or rubbing speed. When this is done with carbon, it is generally found the rubbing speed has *no effect* upon *"wear ratio."* Wear ratio is length of material removed (measured normal to the sliding direction) divided by the length of sliding. It is dimensionless and generally ranges from 10^{-7} to 10^{-12} for carbon sliding on metal when there is no liquid available to form a hydrodynamic film.

D. Shaft Material

The two most important properties for shaft (journal) material are surface hardness and smoothness. The minimum hardness compatible with long life is a Rockwell C value of 35 at the operating temperature. On bearings that operate in the 425 to 650°C (800 to 1200°F) range, coatings such as hard chrome or flame-sprayed ceramics are often used to obtain hardness at high temperatures plus freedom from scale formation.

Carbon bearings can be used with softer shafting. If the bearing is to run dry, a soft carbon low in "polishing action" is required to avoid shaft wear. Bearing wear will be higher under these conditions.

Shaft hardness is not quite so important for bearings that run in grit-free liquids. Under these conditions, part of the load is carried by the hydrodynamic film which forms when the shaft gets up to speed. Even here, though, one must be careful about his choice of carbon grade, treading a path between high shaft wear with a "hard" carbon and high bearing wear with a "soft" carbon. In practice, the choice of carbon grade depends upon the frequency of starts and stops and the bearing load during these transients.

On heavily loaded bearings (above 70 MN/m^2 or 10,000 $lb/in.^2$ projected loads) shaft stiffness is also important. Failure of highly loaded carbon bearings is usually by edge chipping caused by flexure of the shaft.

The shaft finish is more important at high rubbing velocities than at low. Finishes rougher than 1 μm AA (40 μin. AA) can do considerable damage to the carbon bushing before they are worn smooth and coated with carbon wear debris. On the other hand, journal finishes smoother than 0.3 μm (12 μin.) AA are unnecessary, possibly even detrimental to the operation of the bearing.

E. Temperature

Since temperature at the rubbing interface plays a critical role in the formation of a suitable transfer film, the wear rate of many carbon graphites triples over the 90 to 200°C (200 to 400°F) range. Impregnated graphite, however, often shows no increase in wear rates until oxidation (by air) becomes a factor in the 590 to 650°C (1100 to 1200°F) range. Bisson et al.[10] report 100 to 1 variations in wear rates of different carbon grades when tested above 370°C (700°F). Their investigation and the work of other researchers[17] underline the critical importance of selecting the proper carbon grade for operation at high temperature.

Increasing temperature often results in loss of clearance between the shaft and carbon bearing. The coefficient of thermal expansion for most shaft metals is generally two to eight times that of the carbon component. Design procedures to compensate for this differential are discussed later in this chapter.

F. Fluid Environment

Fluid in the bearing clearance removes heat and can provide a hydrodynamic film if rubbing speeds are high enough. When a hydrodynamic film is generated, wear rate becomes negligible and friction drops to a low value. Carbon, impregnated to render it impervious, is an ideal material for hydrodynamic bearings since it will not seize or gall the shaft when it touches the bearing surface during start-up or operational transients. Unlike babbitt, carbon will not soften and melt if the hydrodynamic film is lost. Carbon also has the advantage of excellent corrosion resistance against acids, caustics, and solvents.

Dimensional stability can be a problem for nongraphitized carbons. In the presence of highly polar fluids — such as methyl alcohol, water, and ammonia — these carbons will swell up to 0.0005 mm/mm compared to their air-dried dimensions. Loss of clearance in large diameter, close-tolerance bearings housed in rigid retainers can occur. The proper grade of carbon coupled with its use as a thin-walled sleeve shrunk into metal housings can remedy this problem.

G. Alignment and Runout

Because carbon bearings will not cold-flow to accommodate misalignment or runout of the shaft, localized high loads may lead to hot spots, high bearing wear, and often localized shaft damage. The occurrence of the end chipping and cracking has already been described in Section II Static Load Capacity of Bearings.

Many investigators like Scott[30] have addressed their attention to self-aligning bearing support systems.

H. Bearing Size

Because of their sensitivity to shaft flexure and localized overloads, carbon bearings should have length to diameter (L/D) ratios below 1.0. Higher L/D ratios have been used very successfully for hydrodynamically supported shafts, but in these cases the projected loads were usually below 340 kN/m^2 (50 lb/in.^{2}).

I. Carbon Grade Used

The static load capacity of carbon bearings is generally proportional to the compres-

TABLE 7.1

Guides to the Selection of Carbon Grades for Bearings

Operating condition		
Temperature	Environment	Carbon type
−250 to −35°C (−420 to −30°F)	Liquified gases, vacuum, or dry	Carbon-graphite or graphite with at least 5% inorganic added film former such as silver metal, insoluble fluorides, or phosphates
−35 to 175°C (−30 to 350°F)	Submerged in liquids	Impervious resin-impregnated carbon-graphites; metal-impregnated grades may be used where maximum strength is required
−35 to 175°C (−30 to 350°F)	Dry[a]	Unimpregnated or lightly impregnated carbon-graphite
175 to 350°C (350 to 650°F)	Dry[a]	Unimpregnated carbon-graphite or carbon-graphite lightly impregnated with inorganic film formers
350 to 650°C (650 to 1200°F)	Dry[a]	Graphite impregnated with film former and oxidation inhibitor

[a] Dew point of surrounding air above −20°C.

sive strength of the carbon grade used. Friction and wear rate, however, depend on the interplay of so many factors that they are difficult to predict. If friction and wear rate are critical factors in an application, bearing tests should be conducted. The experience of the carbon supplier is generally extremely useful in selecting appropriate grades. However, when the supplier cannot be consulted, use of Table 7.1 provides some general guidelines for carbon grade selection.

IV. SUBMERGED BEARINGS

A. Hydrodynamic Film Formation

When carbon-lined journal bearings are submerged in a fluid having an appreciable viscosity, self-generation of a hydrodynamic film can greatly reduce friction and wear. Operating conditions which control hydrodynamic film formation are the same for carbon-lined bearings as for other bearings, such as babbitt-lined bearings. Figure 7.17, taken from Shaw and Macks,[31] shows friction as a function of bearing load, speed, L/D ratio, and clearance and fluid viscosity.

The quantitative design of hydrodynamically lubricated journal bearings dates from the work of Tower[32] and Reynolds[33] in the 1880s. Recent step by step design instructions illustrated by examples are provided by O'Connor and Boyd[34] and Rippel.[35]

Figure 7.18 shows both axial and circumferential pressure distribution in the hydrodynamic film. Peak fluid pressures can be in excess of 20 times the projected bearing load, although a more normal range would be three to five times the projected load. End leakage causes the film pressure to drop drastically at the ends of the bearing. The upper curve of Figure 7.17 shows the increase in friction and the reduced load-carrying capacity brought about by end leakage.

Viscosity, of course, plays a major role in load-carrying capacity. Table 7.2 lists the viscosity of a number of fluids spanning five orders of magnitude in viscosity values.

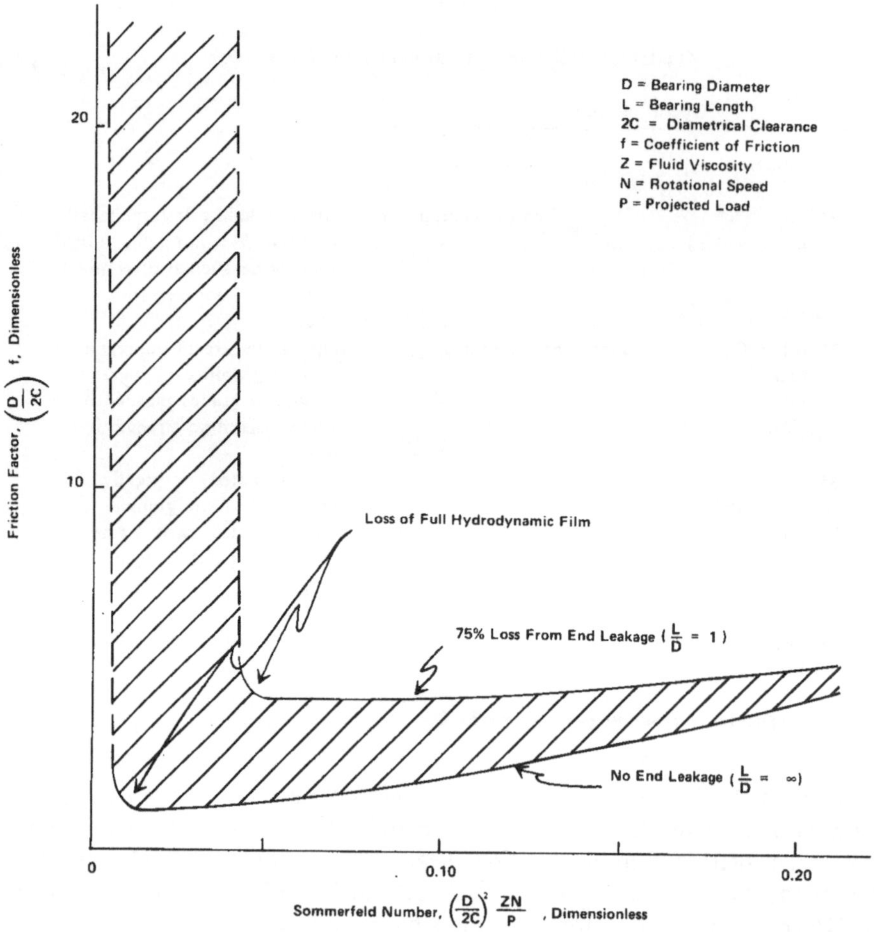

FIGURE 7.17. Analytically determined friction curves for full journal bearing. (From Shaw, M. C. and Macks, E. F., *Analysis and Lubrication of Bearings*, McGraw-Hill, New York, 1949, 240. With permission.)

Note that the viscosity of gases increase as the temperature rises while the viscosity of liquids decreases with rising temperature. Note also that when liquids vaporize, their viscosity drops to roughly 5% of the value of the liquid. This helps explain why vaporization of the liquid in a hydrodynamic film usually destroys its load-carrying capacity.

Carbon-lined bearings are preferable when the hydrodynamic fluid is corrosive, has low viscosity, or is a poor lubricant. Examples of carbon bearing usage will be found in water pumps, fuel pumps handling gasoline, and in air bearings. Because solid-solid contact between the shaft and a carbon bearing does not cause catastrophic failure, carbon-lined bearings can be run at higher temperatures and higher loads (lower Sommerfeld numbers) than babbitt or bronze bearings. Wepfer and Cattabiani[36] ran water-lubricated, carbon-lined bearings successfully for 800 hr at Sommerfeld numbers as low as 0.012. Calculated film thickness ranged from 0.02 mm (0.0008 in.) down to 0.0025 mm (0.001 in.) as their load ranged from 0.13 to 0.55 MN/m² (20 to 80 lb/in.²). Their carbon bearing assemblies, 35 mm (1.38 in.) diameter × 51 mm (2.0 in.) long, showed no measurable wear on either bearing or chrome plated shaft after 800 hr of operation. Other shaft materials gave wear ranging from 0 to 0.012 mm (0.0005 in.) after 800 hr.

A similar test of carbon-lined sleeve bearings was conducted by Dayton et al.[37] Their

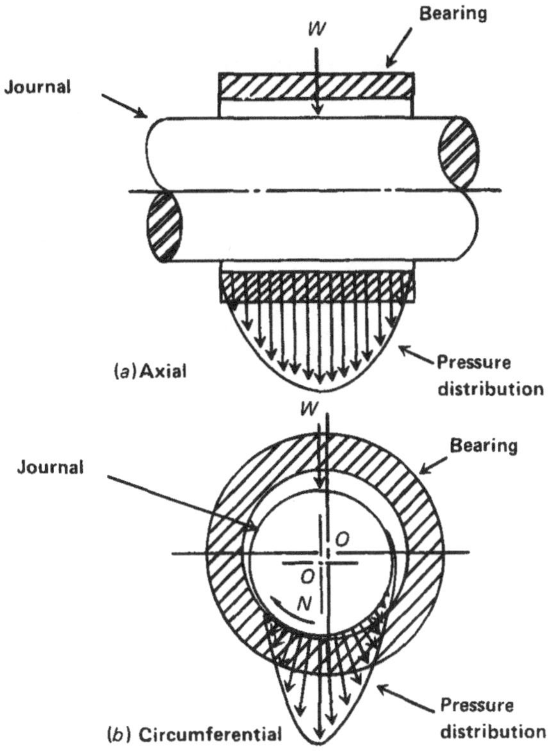

FIGURE 7.18. Pressure distribution of hydrodynamic films.

"lubricant" was 93°C (200°F) water, and their speeds and loads were 3 m/sec (10 ft/sec) and 1.4 MN/m^2 (200 lb/in.^{2}), respectively. They report "promising" results with silver-impregnated carbon liners. Nudelman and Sump[38] confirmed these findings.

When looking at hydrodynamically lubricated carbon bearings, one cannot help but draw comparisons with carbon seal noses. Carbon seal faces operate with film thicknesses an order of magnitude below those found by Wepfer and Cattabiani on bearings. Seal nose wear rates, however, are no higher despite the thinner films. When as much developmental effort has been expended on carbon-lined bearings as has gone into seal design, an order of magnitude improvement in bearing load-carrying capacity will probably result.

V. AIR BEARINGS

Air-lubricated bearings offer low friction, high film stiffness, and a minimum of shaft whirl. Their principle area of application is in spindles for high-speed grinders and other lightly loaded industrial devices and for small gas turbines. Heller et al.[39] describe the design and testing of carbon-lined hydrostatic air bearings. In one application, porous carbon was used on the rubbing surface of the three bearing shoes. Pressurized gas flowed radially inward through the carbon pads and supported the shaft. Frictional power loss at 65,000 rpm was less than 18 W for a 12.7-mm (0.500-in.) journal carrying a radial load of 600 g (1.33 lb). Carbon is an ideal liner for hydrostatic air bearings like these because its gas permeability can be closely controlled and because solid to solid contact during start-up or impulse loads will not gall the shaft.

TABLE 7.2

Viscosity of Selected Fluids

Fluid	State	°F	°C	Viscosity (cP)
Hydrogen	Gas	−431	−257	0.0006
Hydrogen	Gas	73	23	0.009
Ethylene	Gas	73	23	0.010
Hydrogen	Liquid	−431	−257	0.011
Water	Gas	212	100	0.013
Nitrogen	Gas	73	23	0.018
Ammonia	Liquid	73	23	0.11
Nitrogen	Liquid	−314	−192	0.17
Water	Liquid	212	100	0.28
Gasoline	Liquid	73	23	0.3—0.6
Methyl alcohol	Liquid	73	23	0.5
Benzene	Liquid	73	23	0.6
Water	Liquid	73	23	0.94
Kerosene	Liquid	73	23	1—2
Mercury	Liquid	73	23	1.4
Sulfuric acid (100%)	Liquid	73	23	25
Machine oil (light)	Liquid	73	23	60

• 1 cP = 0.001 Nsec/m².

Gas flow through a permeable carbon liner can be calculated using a modification of the Darcy equation:

$$Q_L = \frac{\phi A}{\mu X}\left[\frac{P_2^2 - P_1^2}{2 P_0}\right] \qquad (7.10)$$

where Q_L = volumetric gas flow rate, m³/sec; ϕ = permeability, m² (1 D = 10^{-12} m²); A = area measured normal to gas flow, m²; μ = gas viscosity, Nsec/m² (1 cP = 10^{-3} Nsec/m²); X = carbon thickness measured in direction of flow, m; P_1, P_2 = gas pressures (absolute), N/m²; and P_0 = standard atmosphere = 1.01×10^5 N/m².

This equation assumes that the gas obeys perfect gas laws and that the flow is laminar and isothermal.

Let us illustrate this equation using a carbon sleeve 6 in. long, 3 in. OD, 1 in. ID, having sealed end surfaces and externally pressurized with nitrogen gas at 3 at. The carbon has a permeability of 14,000 μD.

$$\phi = (0.014)(10^{-12}) = 14 \times 10^{-15} \text{ m}^2$$

$$\mu = 0.017 \text{ cP} = 17 \times 10^{-6} \text{ N sec/m}^2$$

$$A_{avg} = \pi \text{ Davg } L = \pi\ (2)\ (6)\ (6.45 \times 10^{-4} \text{ m}^2/\text{in.}^2) = 2.2 \times 10^{-2} \text{ m}^2$$

$$X = 1\ (2.54 \times 10^{-2} \text{ m/in.}) = 2.54 \times 10^{-2} \text{ m}$$

$$P_2 = (3 \text{ atm})\ 1.01 \times 10^5 \text{ N/m}^2 \text{ atm} = 3.03 \times 10^5 \text{ N/m}^2$$

$$P_0 = P_1 = (1 \text{ atm})\ 1.01 \times 10^5 \text{ N/m}^2 \text{ atm} = 1.01 \times 10^5 \text{ N/m}^2$$

$$Q_L = \frac{(14 \times 10^{-15})(2.2 \times 10^{-2})}{(17 \times 10^{-6})(2.54 \times 10^{-2})}\left[\frac{(3.03 \times 10^5)^2 - (1.01 \times 10^5)^2}{2(1.01 \times 10^5)}\right] = 2.9 \times 10^{-4} \text{ m}^3/\text{sec} = 0.80 \text{ ft}^3/\text{min}$$

The gas flow would be somewhat lower in an operating hydrostatic gas bearing be-

TABLE 7.3

Typical Projected Loads for Hydrodynamically Lubricated Journal Bearings

Lubricant	Viscosity (cP)	Load capacity lb/in.^{2}	Load capacity MN/m^2
Oil	1—1000	50—2000	0.35—14
Water	0.2—1	10—100	0.07—0.7
Gas	0.005—0.03	0.5—10	0.003—0.07

cause part of the available pressure is used to support the load. A bearing of this size could be expected to support a 90-kg (200-lb) load.

Table 7.3 shows typical design loads for hydrodynamically lubricated bearings. These values can be increased through the use of a liner material such as carbon in order to permit extremely thin films without the danger of shaft scoring.

VI. DESIGN DETAILS

A. Clearance at Room Temperature

Figure 7.19 gives recommended clearances for carbon-lined bearings. Reduced clearances are recommended for high rotational speeds to reduce shaft whirl.

B. Clearance at Operating Temperature

The thermal expansion rate of carbon is one third to one tenth that of most engineering materials. Figure 7.20 compares thermal expansion values for six common metals with that for a typical mechanical carbon. Because of this differential, the metal shaft expands more rapidly than the carbon bearing when the temperature rises. Loss of clearance thus produced causes more carbon bearing failures than all other factors combined. M.C. Burr[40] and others[41-43] recognized the problem and designed split or multipiece carbon liners to circumvent it. A more common solution is to place the carbon sleeve under radial compression by pressing or shrinking it into a retainer.

C. Press Fitting Carbon Liners

If the peak operating temperature is below 177°C (350° F), carbon liners generally can be press fitted with sufficient interference to avoid liner movement. Interference between the housing inner diameter (ID) and the carbon liner outer diameter (OD) should not exceed 0.15% of the diameter if cracking of the liner during pressing is to be avoided. A 45° chamfer on the liner OD reduces end chipping and aids in aligning the bearing with the housing. Before beginning the pressing operation, both the liner and the housing should be carefully inspected for roundness and uniform diameter along their axes. Burrs and sharp edges on the housing should be removed to avoid shaving the softer carbon liner during the press.

The following procedure should be used for pressing:

(1) Align the carbon liner so that it is coaxial with the hole.
(2) Distribute the pressing force across the entire end of the bearing to avoid localized crushing or chipping.
(3) Once the press is started, the force and speed of press should continue without pause until the bearing is fully seated.
(4) On large bearings the housing may be lightly coated with a lubricant such as white lead in oil to reduce the pressing force.

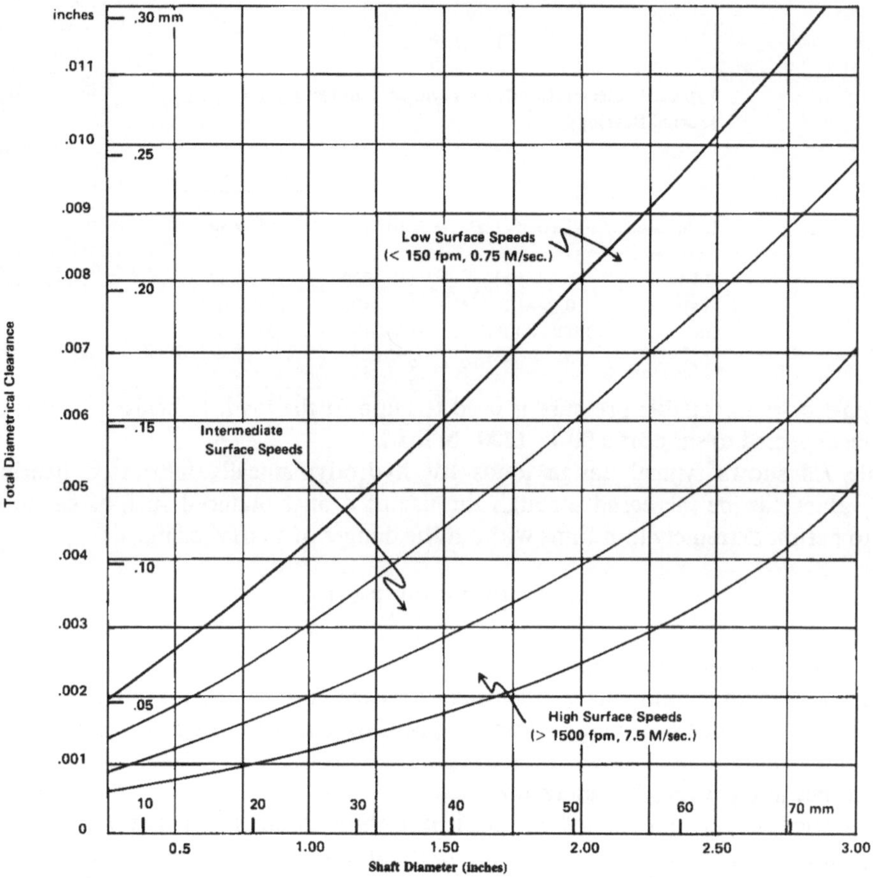

FIGURE 7.19. Clearance for carbon-lined bushings. Arrows refer to zones, not any one curve.

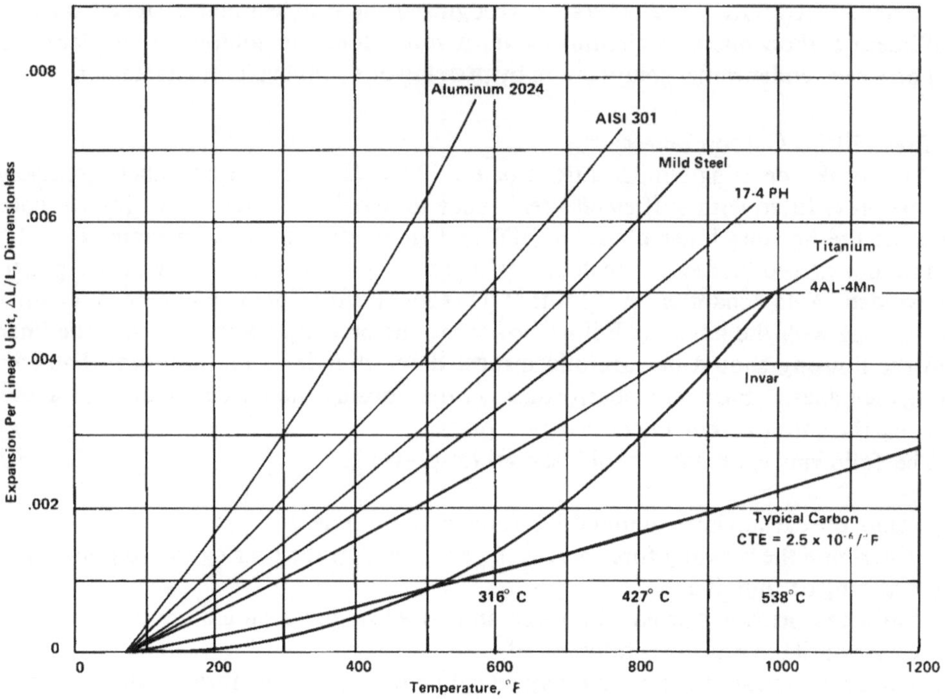

FIGURE 7.20. Thermal expansion of carbon and other engineering materials.

D. Shrink Fitting

Shrink fitting coupled with thin-walled carbon liners makes constant clearance operation possible over a wide temperature range. Highly compressed carbon liners will not yield and relax; rather, they will press radially against the housing and faithfully follow its thermally induced expansion. If the housing has the same expansion rate as the shaft, a constant clearance between bearing and shaft results.

E. Sizing Liners for Shrink Fits

The following formula assumes that radial thickness of both housing (A) and liner (B) are small (less than 10%) compared to their diameter and that their axial lengths are equal:

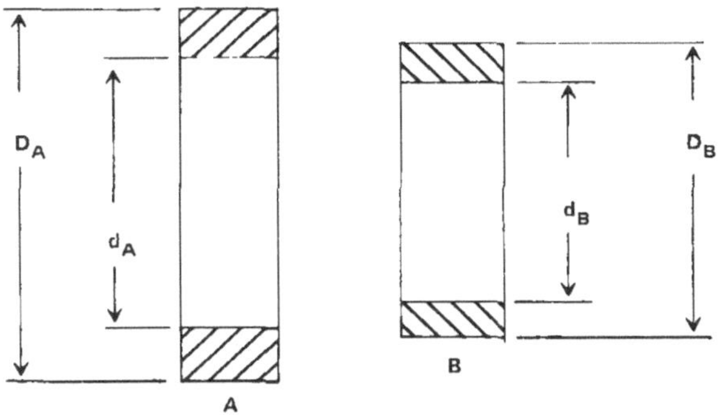

Ring A = Metal Housing
Ring B = Carbon liner
E_A = Tensile modulus of A
E_B = Compressive modulus of B
α A = Tensile stress in A
α B = Compressive stress in B

The room temperature diametrical interference, S_{AB}, is

$$s_{AB} = D_B - d_A \tag{7.11}$$

The tensile stress in the outer, metal housing is

$$\sigma_A = s_{AB} E_A \Big/ \left[1 + \frac{(D_A - d_A) E_A D_B}{(D_B - d_B) E_B d_A} \right] d_A \tag{7.12}$$

The compressive stress on the inner carbon liner is

$$\sigma_B = s_{AB} E_B \Big/ \left[1 + \frac{(D_B - d_B) E_B d_A}{(D_A - d_A) E_A D_B} \right] D_B \tag{7.13}$$

Since D_B is essentially the same value as d_A, these equations simplify to

$$\sigma_A = \frac{s_{AB} E_A}{d_A} \Big/ \left[1 + \frac{(D_A - d_A) E_A}{(D_B - d_B) E_B} \right] \tag{7.14}$$

$$\sigma_B = \frac{s_{AB} E_B}{D_B} \Big/ \left[1 + \frac{(D_B - d_B) E_B}{(D_A - d_A) E_A} \right] \tag{7.15}$$

As an example, let housing A be 310 stainless steel having a room temperature tensile (yield) strength of 340 MN/m² (50,000 psi) and a modulus of 200 GN/m² (30 × 10⁶ psi). Liner B will be carbon P-5Ag (see Table 4.1) which has a compressive strength of 310 MN/m² (45,000 psi) and a modulus of 26 GN/m² (3.8 × 10⁶ psi):

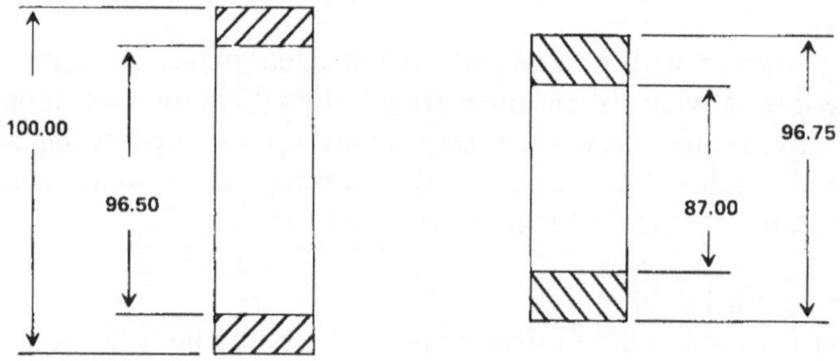

$$S_{AB} = 96.75 - 96.50 = 0.25 \text{ mm}$$

$$\sigma_A = \frac{(0.25 \times 10^{-3})(200 \times 10^9)}{96.5 \times 10^{-3}} \Bigg/ \left[1 + \frac{(100.00 - 96.50) \times 10^{-3}\ (200 \times 10^9)}{(96.75 - 87.00) \times 10^{-3}\ (26 \times 10^9)}\right] = 138 \text{ MN/m}^2$$

or 20,000 psi tensile hoop stress in metal retainer

$$\sigma_B = \frac{(0.25 \times 10^{-3})(26 \times 10^9)}{96.75 \times 10^{-3}} \Bigg/ \left[1 + \frac{(96.75 - 87.00) \times 10^{-3}\ (26 \times 10^9)}{(100.00 - 96.50) \times 10^{-3}\ (200 \times 10^9)}\right] = 49.3 \text{ MN/m}^2$$

or 7150 psi compressive stress in the carbon liner. After assembly the retainer outer diameter will be

$$D_{A_f} = D_A\left[1 + \frac{\sigma_A}{E_A}\right] = 100.00\,(1 + 0.138/200) \cong 100.07 \text{ mm}$$

and the liner bore will be

$$d_{B_f} = d_B\left[1 - \frac{\sigma_B}{E_B}\right] = 87.00\,(1 - 0.0493/26) \cong 86.83$$

Axial distortion (Poisson's effect) will reduce the change in diameters slightly, but the effect is negligible at these strain levels.

F. Dimensional Changes After Shrink Fits

Figures 7.21 and 7.22 provide a simplified way of finding dimensional changes of the housing and liner resulting from shrink fits. These simplified relationships assume the wall thickness of both retainers and liner are small (less than 10%) compared to the diameters:

$$\left[(D_A - d_A)/2D_A\right] < 0.1$$

and

$$\left[(D_B - d_B)/2D_B\right] < 0.1$$

They also assume both housing and liner exhibit linear stress-strain relations.

G. Stress Resulting from Shrink Fits

Figure 7.23 shows the stress in the housing or liner resulting from the press or shrink fit. This stress must always be below the yield point of the materials.

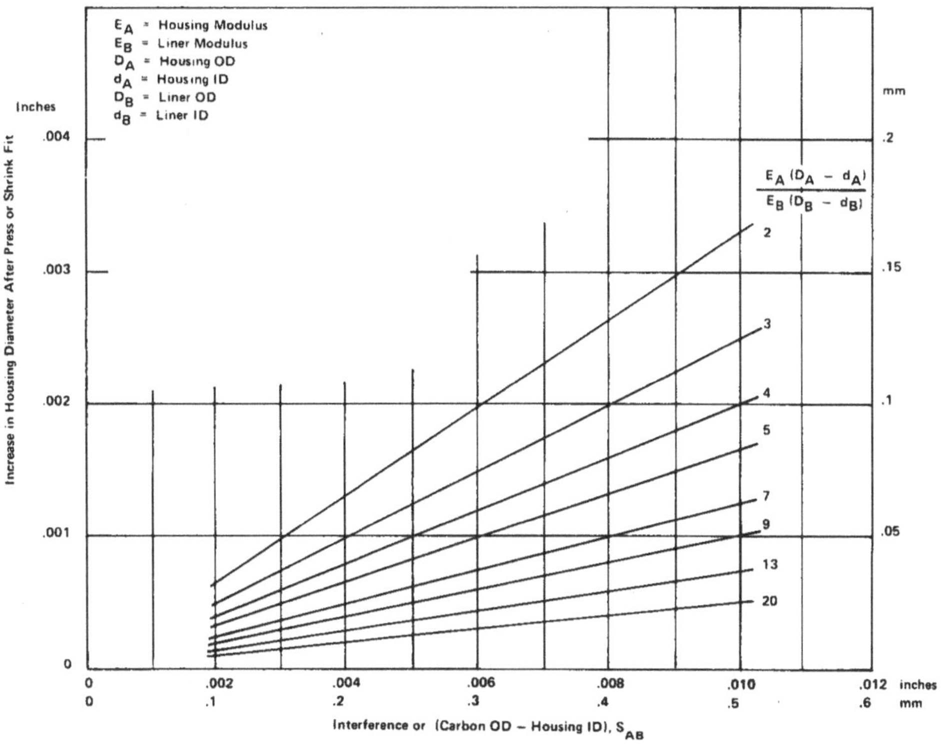

FIGURE 7.21. Increase in housing diameter after press or shrink fit of carbon liner.

H. Example of Use of Figures to Design Shrink Fits

Using dimensions of the prior example:

$$E_A = 200\ \text{GN/m}^2\ (30 \times 10^6\ \text{lb/in.}^2)$$

$$D_a - d_a = 100.00 - 96.50 = 3.50\ \text{mm}\ (0.138\ \text{in.})$$

$$E_B = 26\ \text{GN/m}^2\ (3.8 \times 10^6\ \text{lb/in.}^2)$$

$$D_B - d_B = 96.75 - 87.00 = 9.75\ \text{mm}$$

$$E_A (D_A - d_a) / E_B (D_B - d_B) = \frac{200}{26} \frac{(3.5)}{(9.75)} = 2.8$$

$$S_{AB} = 96.75 - 96.50 = 0.25$$

From Figure 7.21, increase in housing diameter = 0.07 mm;
from Figure 7.22, decrease in liner bore = 0.17 mm.
The stress level, in turn, depends on modulus, average diameter, and change in diameter. For housing,

$$E/d \times 10^6 = 200{,}000/0.10 \times 10^6 = 2\ \text{MN/m}^3$$

Change in diameter = 0.07.
From Figure 7.23, housing stress ≅ 138 MN/m² (20,000 lb/in.²) in tension. For liner

$$E/d \times 10^6 = 26{,}000/0.09 \times 10^6 = 0.3\ \text{MN/m}^3$$

Change in diameter = 0.17 mm.

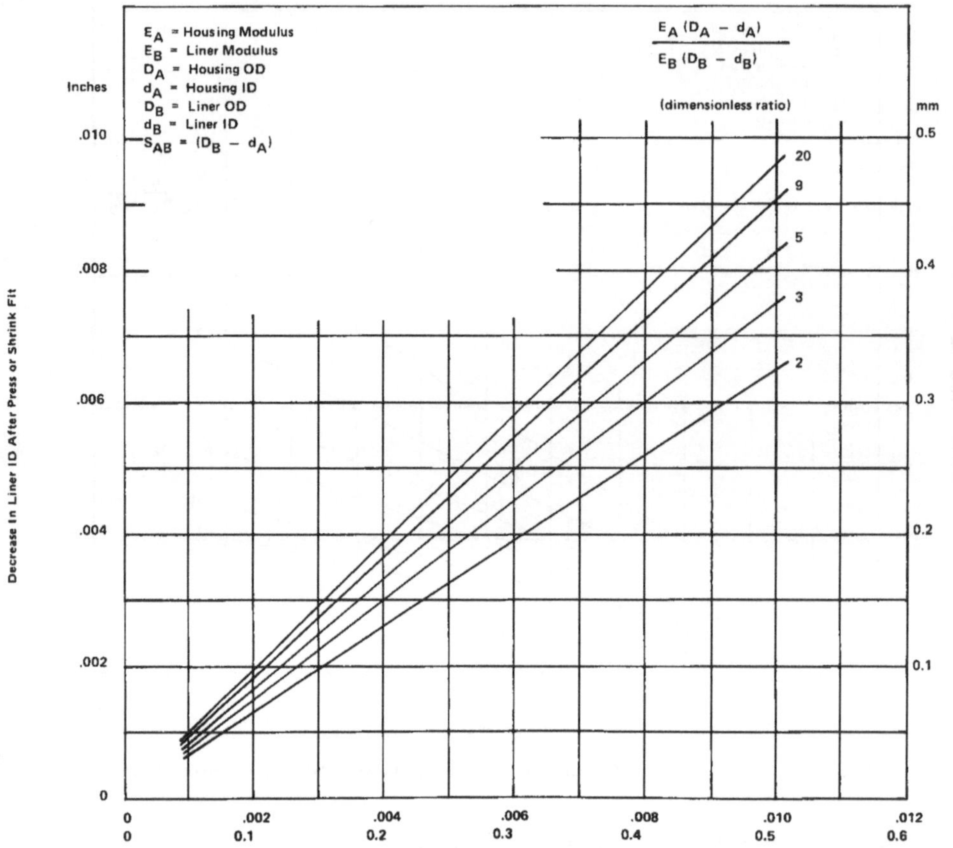

FIGURE 7.22. Decrease in liner diameter after press or shrink fit of carbon liner.

From Figure 7.23, liner stress ≅ 50 MN/m² in compression.

Since both stress levels are well below the elastic limits of the materials, the design is a safe one.

I. Cementing Liners in Housings

If the operating temperature is below 127°C (260°F), it is often feasible to cement the liner in the housing. Epoxy resins are commonly used. A major drawback to this approach is the possibility that the cement will pull loose from either housing or liner surface and permit the liner to shift. Other difficulties arise from poor thermal contact between liner and housing and high friction caused by contamination of the rubbing surface with the cement. Despite these hazards, cemented carbon liners are successfully used for many light duty applications.

1. Heat Generation in Dry Carbon Bearings

Before deciding upon cement to hold the liner in place, one should estimate bearing operating temperature. Heat removal rather than wear rate is usually the biggest obstacle to running carbon bearings at high PV levels. Consider the following bearing:

Shaft diameter	25.3 mm	(0.996 in.)
Shaft density	7.7 g/cc	(7.7 g/cc)
Bore diameter	25.4 mm	(1.00 in.)
Carbon liner thickness (radial)	1.27 mm	(0.05 in.)
Carbon liner density	1.8 g/cc	(1.8 g/cc)
Bearing length	25.4 mm	(1.00 in.)

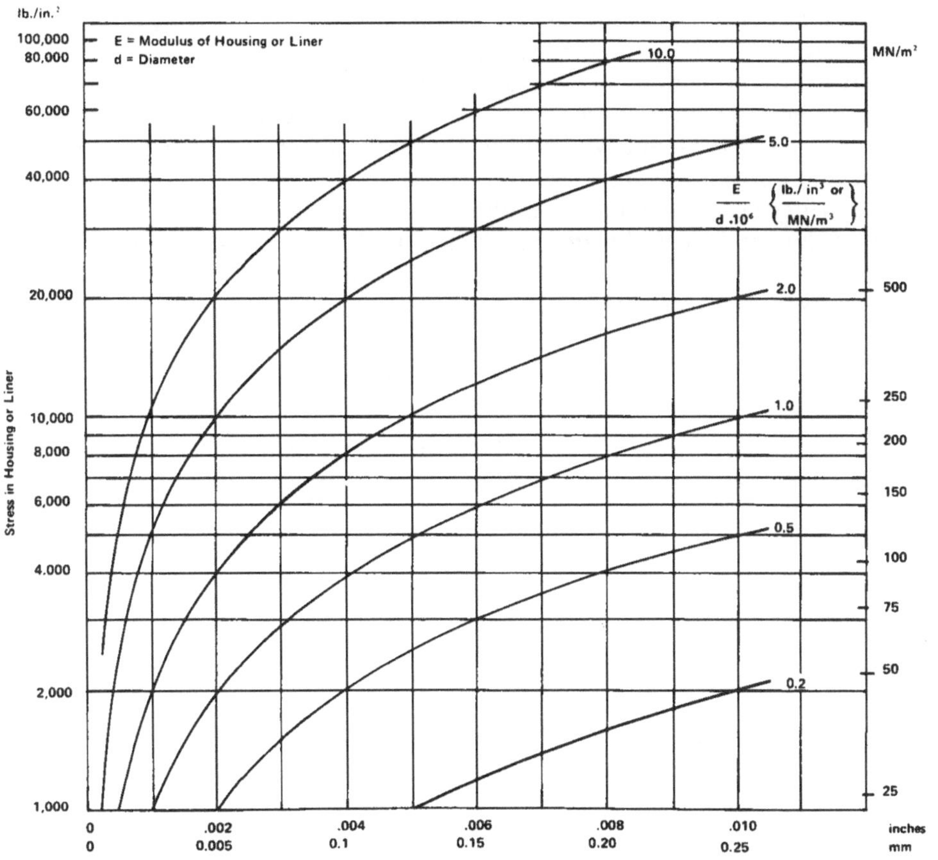

FIGURE 7.23. Stress in housing or liner after press or shrink fit.

Load	4448 N	(1000 lb_f)
Speed	1 rps	(60 rpm)
Coefficient of friction	0.05	(0.05)
Conversion factor	1 N·m/W sec	(778 ft lb/Btu)

Heat generation rate:

$$[\ (4448)\ (0.05)\ ]\ \ [\pi\,(0.0253)\ ]\ \ 60\ \text{rpm}/60\ =\ 17.6\ \text{W}\ (1\ \text{Btu/min})$$

Weight of carbon liner:

$$\frac{\pi}{4}\ (2.794^2 - 2.540^2)\ 2.54\ \times\ \frac{1.8}{\text{cm}^3}\ \text{g}\ =\ 4.86\ \text{g}$$

Using a liner heat capacity of 1.88 J/g°C (0.45 Btu/lb °F) and assuming half the heat goes into the bearing, the rate of carbon temperature rise is

$$(17.6/2)\ /4.86\ \times\ 1.88\ =\ 0.96°\ \text{C/sec}\ (104°\text{F/min})$$

Using a shaft heat capacity of 0.38 J/g °C (0.09 Btu/lb °F) and assigning the rest of the heat to the shaft, its rate of temperature rise becomes

$$(17.6/2)/\ (\ \pi/4)\ (2.53^2)\ (2.54)\ (7.7)\ \times\ 0.38\ =\ 0.23°\text{C/sec}\ (26°\text{F/min})$$

A temperature gradient will become established quickly with heat flowing from the liner to the shaft. Heat leaves the bearing area by flowing axially along the shaft and radially through the bearing support structure.

From the above calculations, it is obvious that adequate heat removal is essential if bearing operating temperatures are to stay within reasonable limits. It is also obvious that insertion of a low conductivity cement between liner and housing can markedly increase bearing operating temperature.

To get a better perspective on the problem, consider the above described carbon bearing to be housed in a steel plate 25 mm (1 in.) thick. Heat will flow radially outward from the bearing at a rate of 8.8 W (0.5 Btu/min), assuming half the heat is carried away by the shaft and its appendages:

$$\frac{dq}{d\theta} = k\,A\,\frac{dt}{dr} = 8.8 \text{ W (30 Btu/hr)}$$

where k = metal thermal conductivity ≅ 45 W/m°C (26 Btu/hr °F/t); A = area normal to heat flux = π r(0254) = 0.0798r; and dt/dr = radial temperature gradient, °C/m. Solving for dt/dr:

$$dt/dr = 8.8/45\ (0.0798r) = 2.45/r$$

At the outer surface of the liner, whose *diameter* is 27.9 mm,

$$dt/dr = 2.45/\ (0.0279)(1/2) = 175°\text{C/m} = 0.17\ °\text{F/mm} \cong 8°\text{F/in.}$$

Thus, it is apparent that if there is no thermally insulating barrier between bearing and housing and if a proper heat sink is provided, the temperature rise of the bearing and its immediate environment would not lead to serious problems.

The case described above is a mild one with a PV of but 0.55 MN/m² × m/sec (15,700 lb/in.² × ft/min) and with a bearing housing having good thermal conductivity. At higher loads and speeds, or with a housing less favorable for heat dissipation, the temperature of the bearing can rise hundreds of degrees above ambient before equilibrium is established between rate of heat generation and rate of heat lost by conduction plus convection plus radiation.

As the temperature rises, shaft, bearing liner, and housing expand. Clearance must not be lost between shaft and bearing liner lest the bearing seize up. On the other hand, interference must be retained between liner and housing lest the liner shift axially or start to rotate in the housing. Interference between liner and housing must not drop below 0.0005 in. per inch diameter if this is to be avoided.

Table 7.4 lists approximate values for the temperature rise in dry running carbon bearings. The temperature rise values assume unrestricted natural convection of air in the bearing area plus good thermal contact between bearing and its support.

It should be noted that Table 7.4 is only a guide. Some shaft and carbon-grade combinations have friction coefficients half those of others. For some, the friction rises with increasing temperatures, whereas for others it falls. Unless the designer has considerable experience with carbon bearings, it is not likely that he will choose the best carbon grade for his applications. Most carbon bearing manufacturers offer free design assistance to aid in selection of the best grade for the application at hand.

2. Illustrative Design

The preceding examples have considered certain aspects of designing carbon bearings. Now we will go through the complete design of a metal jacketed bearing from beginning to end, noting approximations and compromises as we go.

TABLE 7.4

Temperature Rise Due to Frictional Heating in Dry Bearings

Load factor (PV)		Approximate temperature rise	
lb ft / in.² min	N M / m² sec	°F	°C
10,000	0.35	100	55
20,000	0.70	175	97
30,000	1.05	240	133
40,000	1.40	300	167
50,000	1.75	350	195
60,000	2.10	400	222
80,000	2.80	500	278
100,000	3.50	600	333

Note: Assumes (1) unimpeded natural convection across bearing; (2) good thermal contact with a metal support structure; and (3) black body radiation to surroundings at a uniform ambient temperature.

Given data (room temperature dimensions)

Shaft diameter (mm)	$25.000\,^{+0.000}_{-0.013}$
Shaft material	17-4 Ph stainless steel
Shaft finish (μm)	0.5 center line average (c.l.a.)
Shaft hardness (R_c)	50
Shaft speed (rpm)	1700
Load (steady state) (kg force)	50
Ambient temperature (°C)	175
Ambient fluid	Oven gases containing 3% CO_2 1% Co, 2% H_2O, 16% O_2, N_2 rem.
Housing bore (mm)	$37.500\,^{+0.025}_{-0.000}$
Housing thickness (mm)	25.0

a. Design Steps

1. Calculate PV —

$$[\,(50 \times 9.807) / (0.025)(0.025)\,]\;[\,(1700/60)\,]\;[\,(\pi)(0.025)\,] = 1.75\ (\text{MN/m}^2)\ (\text{m/sec}) \text{ or } 50{,}000\ (\text{lb/in.}^2)\ (\text{ft/min})$$

2. Estimate Temperature Rise — Lacking data, we will assume this application meets the criteria of Table 7.4 and estimate the temperature rise to be 195 °C.

3. Choose carbon grade — This is the key step. Estimated bearing temperature is 370°C (698°F), 175 + 195°C. According to Table 7.1, this is near the border between an unimpregnated carbon-graphite and an impregnated graphite. The loads are light, and we do not need the strength of a carbon-graphite; hence, a graphite is recommended. When lightly impregnated with an appropriate film former, its friction coefficient will be lower and its thermal conductivity higher than a carbon-graphite. The selected grade has properties as follows:

Compressive strength (MN/m^2)	138 (20,000 psi)
Modulus of elasticity (GN/m^2)	13.8 (2.0×10^6 psi)
Coefficient of thermal exp. ($°C^{-1}$)	3.8×10^{-6} ($2.1 \times 10^{-6}/°F$)

4. Choose retainer material — The retainer or jacket should have excellent corrosion resistance at the peak operating temperature and a thermal expansion coefficient close to that of the shaft material. A 310 stainless steel is recommended for this case.

Yield (tensile) strength (MN/m^2)	345 (50,000 psi)
Modulus of elasticity (GN/m^2)	207 (30×10^6 psi)
Coefficient of thermal exp. ($°C^{-1}$)	16×10^{-6} ($9 \times 10^{-6}/°F$)

5. Housing bore at peak temperature (370°C) —

$$[\,37.500 \,{}^{+0.025}_{-0.000}\,]\,[\,1 + (370 - 23)\,(20 \times 10^{-6})\,] = 37.760 \,{}^{+0.025}_{-0.000} \text{ mm}$$

where 20×10^{-6} is the thermal expansion coefficient of the steel support structure. Neglect the radial compressive effect on the bore caused by the fact that the metal at some distance from the bore is cooler.

6. Select bearing jacket OD at peak temperature (370°C) — To be sure jacket does not rotate in housing bore, an interference of 0.0005 mm/mm is recommended:

$$(37.760 + 0.025)\,(1.0005) = 37.804 \text{ mm}$$

Add tolerance and the jacket OD becomes $37.804 \,{}^{+0.025}_{-0.000}$ mm at 370°C.

7. Select jacket ID at peak temperature (370°C) — There is little benefit from carbon liners thicker than 1.33mm measured radially; hence, we can arbitrarily set hot jacket ID at, say, 2.6 mm greater than the cold diameter, say $27.600 \,{}^{+0.025}_{-0.000}$ mm.

8. Select carbon liner OD at peak temperature (370°C) — An interference of 0.002 mm/mm is ample to ensure there will be no liner shifting. Thus, liner OD at 370°C should be

$$(27.600 + 0.025)\,(1.002) = 27.680 \,{}^{+0.025}_{-0.000} \text{ mm}$$

9. Calculate shaft diameter at peak temperature (370°C) —
From Figure 7.20, $\Delta L/L$ of 17-4 stainless steel is 0.0042; hence, shaft diameter at 370°C becomes

$$(25.00 \,{}^{+0.000}_{-0.013})\,(1.0042) = 25.105 \,{}^{+0.000}_{-0.013}$$

10. Select clearance at peak temperature (370°C) — Rotational speed is 1700 rpm giving a rubbing speed of

$$[\,(1700/60)\,][\,(0.025\,\pi)\,] = 2.23 \text{ m/s}$$

From Figure 7.19, clearance should be 0.07 mm.

11. Calculate liner ID at peak temperature (370°C) —

$$(25.105 \,{}^{+0.000}_{-0.013}) + 0.07 = 25.175 \,{}^{+0}_{-0.013}$$

12. Summarize the first 11 steps —

	Dimensions (mm) At 23°C	Dimensions (mm) At 370°C
Shaft diameter	$25.000 \begin{smallmatrix} +0.000 \\ -0.013 \end{smallmatrix}$	$25.105 \begin{smallmatrix} +0.000 \\ -0.013 \end{smallmatrix}$
Carbon liner ID (when in housing)		25.175 ± 0.013
Carbon liner OD (not housed)		$27.680 \begin{smallmatrix} +0.025 \\ -0.000 \end{smallmatrix}$
Metal jacket ID (not housed)		$27.600 \begin{smallmatrix} +0.025 \\ -0.000 \end{smallmatrix}$
Metal jacket OD (containing liner)		$37.804 \begin{smallmatrix} +0.025 \\ -0.000 \end{smallmatrix}$
Support bore	$37.500 \begin{smallmatrix} +0.025 \\ -0.000 \end{smallmatrix}$	$37.760 \begin{smallmatrix} +0.025 \\ -0.000 \end{smallmatrix}$
Clearance, shaft to liner		0.057 to 0.096
Interference, carbon OD to jacket ID		0.055 to 0.105

13. Calculate stretch of the metal jacket caused by 0.080-mm interference from carbon liner at 370°C — For Figure 7.21 and neglecting any temperature effect on modulus:

$$\frac{E_A (D_A - d_A)}{E_B (D_B - d_B)} = \frac{207\ (37.8 - 27.6)}{13.8\ (27.68 - 25.17)} = 61$$

Interference, S_{AB} = 0.08; hence, housing stretch (from Figure 7.21) will be less than 0.005 mm, and substantially all of the interference will be seen as as a close-in of the carbon ID after it is fitted into its jacket. This will be recalled when we size the carbon liner.

14. Calculate room temperature OD of metal jacket —

$$(37.804 \begin{smallmatrix} +\ 0.025 \\ -\ 0.000 \end{smallmatrix})\ [1 - (370-23)\ 16 \times 10^{-6}] = 37.594 \begin{smallmatrix} +\ 0.025 \\ -\ 0.000 \end{smallmatrix} \text{ mm}$$

15. Calculate room temperature ID of metal jacket —

$$(27.600 \begin{smallmatrix} +\ 0.025 \\ -\ 0.000 \end{smallmatrix})\ [1 - (370-23)\ 16 \times 10^{-6}] = 27.447 \begin{smallmatrix} +\ 0.025 \\ -\ 0.000 \end{smallmatrix} \text{ mm}$$

16. Calculate carbon liner OD at room temperature —

$$(27.680 \begin{smallmatrix} +\ 0.025 \\ -\ 0.000 \end{smallmatrix})\ [1 - (370-23)\ 3.8 \times 10^{-6}] = 27.644 \begin{smallmatrix} +\ 0.025 \\ -\ 0.000 \end{smallmatrix} \text{ mm}$$

17. Calculate room temperature interference between carbon liner OD and metal jacket ID —

$$(27.644 \begin{smallmatrix} +\ 0.025 \\ -\ 0.000 \end{smallmatrix}) - (27.447 \begin{smallmatrix} +\ 0.025 \\ -\ 0.000 \end{smallmatrix}) = 0.222 \text{ to } 0.172 \text{ mm}$$

18. Calculate stretch of metal jacket caused by 0.200-mm interference between carbon liner OD and jacket ID at 23°C —

$$\frac{E_A (D_A - d_A)}{E_B (D_B - d_B)} = 61$$

From Figure 7.21, stretch = 0.010 mm.

19. Close-In of liner ID at 23°C — As shown above, most of the interference will show up as a reduction in carbon ID and OD.

20. Calculate compressive stress on carbon liner — Using equation 7.15.

$$\sigma_B = \frac{(2.0 \times 10^{-4})\ (13.8 \times 10^9)}{0.026} \Bigg/ \left[1 + \frac{(0.0025)\ 13.8 \times 10^9}{(0.0102\ (207 \times 10^9}\right] = 104\ \frac{MN}{m^2}$$

This is a high compressive stress, but still within the 138 MN/m^2 capability of the carbon. One might take a chance on losing hot interference and reduce this stress level 25% by reducing the carbon OD by 0.050 mm. This would reduce breakage losses during assembly and thereby reduce part cost. The risk of losing hot interference will be mitigated by compression of the jacket when it is pressed into the housing bore. This reduction has been made in the table which follows.

21. Summarize the first 20 steps —

	Dimensions (mm)	
	At 23°C	At 370°C
Shaft diameter	25.000 + 0.000 − 0.013	25.105 + 0.000 − 0.013
Carbon liner ID (in housing)		25.175 ± 0.013[a]
Carbon liner ID (free)		25.205 ± 0.013[a]
Carbon liner OD (free)	27.594 + 0.025 − 0.000	27.630 + 0.025 − 0.000
Metal jacket ID (with liner)	27.447 + 0.025 − 0.000	27.600 + 0.025 − 0.000
Metal jacket ID (free)	27.440 + 0.025 − 0.000	27.600 + 0.025 − 0.000
Metal jacket OD (free)	37.587 + 0.025 − 0.000	37.804 + 0.025 − 0.000
Metal jacket OD (with liner)	37.594 + 0.025 − 0.000	37.804 + 0.025 − 0.000
Support bore	37.500 + 0.025 − 0.000	37.760 + 0.025 − 0.000
Clearance, shaft to liner		0.057 to 0.096[a]
Interference, carbon OD to jacket ID	0.129 to 0.179	0.005 to 0.055
Interference, jacket OD to support bore	0.069 to 0.119	0.019 to 0.069

[a] Increased in steps 22 to 24.

Note that at 370°C the metal jacket dimensions are not affected by the carbon liner but that the liner is compressed a nominal 0.030 mm by being placed in the jacket.

22. Calculation of carbon liner ID (Free) at 23°C —

$$(25.205 \pm 0.013)\ [\,1 - (370 - 23)\ 3.8 \times 10^{-6}\,] = 25.172 \pm 0.013.$$

23. Calculation of carbon liner ID after installation in the metal jacket — Room temperature interference liner OD to jacket ID is 0.154 mm nominal reduced by 0.007 by metal stretch. All of the residual interference will be reflected as a reduced ID. This installed room temperature liner ID will be

$$25.172 \pm 0.013 - (0.154 - 0.007) = 25.025 \pm 0.013.$$

24. Examination of room temperature clearance — This clearance will range from 0.012 to 0.051 mm, not enough to be safe. We should remove another 0.072 mm from the carbon ID even though this runs the diametrical clearance up to 0.129/0.168 at 370°C. A second reason for this much added clearance is radial compression of the entire jacketed assembly when it is pressed into the housing bore. It is difficult to estimate how interferences of this type will be distributed between the support structure

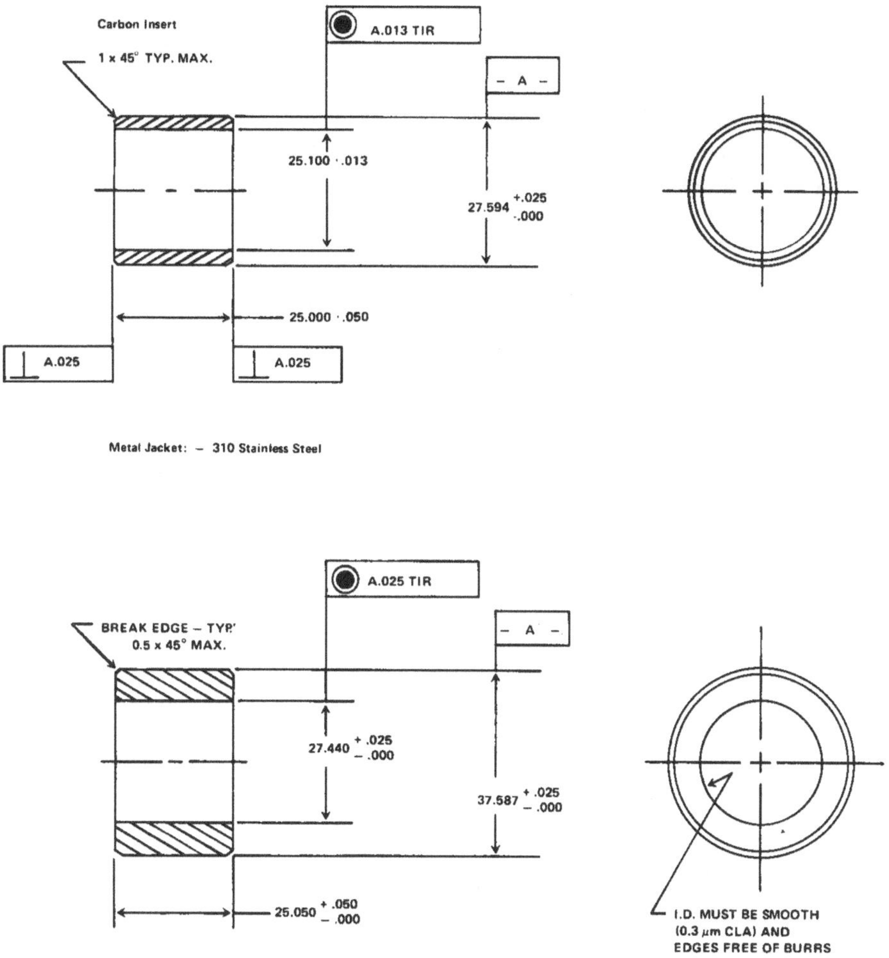

FIGURE 7.24. Drawing of parts in illustrative design.

and the bearing insert, but half or more must usually be accepted by the pressed in bearing assembly. Figure 7.24 is a drawing of the final assembly giving dimensions of each component before it is assembled.

25. Shrink fitting carbon liner into metal jacket — Assume we want a 0.025-mm clearance for the shrink fit, jacket must be heated until its ID is

$$27.594 + 0.025 + 0.025 = 27.644 \text{ mm}$$

$$T = (27.644 - 27.440) / (27.440)\ (16 \times 10^{-6}) = 465°C$$

Retainer will need to be heated to 490°C so that the carbon liner (at 23°C) will slip inside it. This temperature will not alter the metallurgy of the metal or cause oxidation, provided the exposure time is under 15 min.

This example shows some of the compromises that enter into the design of carbon lined bearings as well as illustrating the details of calculation.

VII. INTERNAL GROOVING

Carbon bearings intended for use in dusty environments or grit laden liquids usually contain grooves on the inner surface of the carbon liner. Straight (axial) grooves are cheaper to produce but require care in installation to be sure the grooves are not in the load-carrying area. Helical grooves are used when it is not feasible to control orientation of the carbon sleeve.

For installation where a hydrodynamic film between shaft and bore is sought, groove location and shape become as important as clearance in controlling performance. This topic is well covered in O'Connor and Boyd[34] and Rippel.[35]

VIII. PROBLEM ANALYSIS

This section is added as an aid in searching for cause of bearing failures.

1. Seizure during start-up
   A. Error in dimension or tolerances leaving insufficient clearance
   B. Misalignment during assembly
   C. Foreign material lodged in clearance
   D. Exudation or corrosion products in clearance
2. Seizure midway in expected life
   A. Overload has bent shaft or changed alignment
   B. Temperature excursion has closed clearance
   C. Wear in other parts of system has changed alignment
   D. Wear debris lodged in clearance
3. Running Hot
   A. Wrong carbon grade used
   B. All factors which are shown above for "seizure"
   C. Poor shaft finish or corroded shaft
   D. Shaft too soft
   E. Bearing liner loose in housing
4. Squealing
   A. Wrong carbon grade used causing overfilming
   B. Bearing has been contaminated with grease
   C. Poor shaft finish
   D. Load too light (operating temperature too low)
5. Pitting or corrosion of shaft
   A. Stray electric currents
   B. Wrong shaft material for environment
   C. Galvanic action
   D. Bleed out of oxidation inhibitor from graphite bearing
6. Short life due to high wear
   A. Wrong carbon grade used
   B. Improper shaft material — or finish
   C. Load higher than design
   D. Temperature higher than design
   E. Carbon liner cracked during installation
   F. Carbon liner became loose in housing and rotated with shaft

REFERENCES

1. **Kellogg, L. G. and Dewart, W. G.**, SNAP 8 reactor bearing development, *ASLE Trans.*, 9, 325, 1966.
2. **Carson, R. W.**, Selecting high temperature sleeve bearings, *Prod. Eng., N.Y.*, January 4, 1965.
3. **Paxton, R. R. and Scilingo, A. A.**, Carbon-graphite bearings for high loads, *Lubr. Eng.*, 25, 246, 1969.
4. **Johnson, R. L., Swikert, M. A., and Bailey, J. M.**, *Wear of Typical Carbon-Base Sliding Seal Materials for Temperatures to 700°F*, NACA Technical Note 3595, 1956.
5. **Rabinowicz, E.**, *Friction and Wear at Elevated Temperatures*, WADC Tech. Rep. 59-603, 1960.
6. **Rabinowicz, E. and Imai, M.**, *Friction and Wear at Elevated Temperatures*, WADC Tech. Rep. 59-603, Part IV, July 1962.
7. **Sibley, L. B., Allen, C. M., Zielenbach, W. J., Peterson, C. L., and Goldthwaite, W. H.**, *A Study of Refractory Materials for Seal and Bearing Applications in Aircraft Accessory Units and Rocket Motors*, WADC Tech Rep. 58-299, 1958.
8. **Swikert, M. A. and Johnson, R. L.**, Wear of carbon-type seal materials with varied graphite content, *ASLE Trans.*, 1, 115, 1958.
9. **Lotos, E. J.**,*Mechanism and Wear of Non-Metallic Materials*, WADC Tech. Rep. 59-316, 1959.
10. **Bisson, E. E., Johnson, R. L., and Anderson, W. J.**, Friction and lubrication with a solid lubricant at temperatures to 1000°F with particular reference to graphite, Paper No. 23, in Institute of Mechanical Engineering Conf. Lubrication and Wear, October 1957.
11. **Driesner, A. R. and Wagner, P.**, *J. Appl. Phys.*, 29, 901, 1958.
12. **Wagner, P. and Driesner, A. R.**, *Friction Coefficients of Graphite at Extreme Temperatures*, No. LA-2146, U.S. Atomic Energy Commission, 1958.
13. **Earp, F. K. and Hill, M. W.**, *Industrial Carbon and Graphite Papers Conf., London*, 1958, 326.
14. **Carlson, R. K. and Schwemer, W. C.**, U.S. Patent 3,029,167, 1962.
15. **McKee, D. W.**, Effect of adsorbed phosphorus oxychloride on the oxidation behavior of graphite, *Carbon*, 10, 491, 1972.
16. **Nickerson, J. D.**, U.S. Patent 2,906,632, 1959.
17. **Paxton, R. R. and Shobert, W. R.**, Testing carbon for seals and bearings, *Lubr. Eng.*, 17, 27, 1961.
18. **Shobert, W. R. and Olson, W. O.**, Friction and wear for carbon bearings, *Prod. Eng.*, N.Y., November 8, 1965.
19. **Wisander, D. W., Hady, W. F., and Johnson, R. L.**, *Friction Studies of Various Materials in Liquid Nitrogen*, NACA Tech. Note 4211, 1958.
20. **Bisson, E. E.**, Friction of metals, lubricating coatings and carbon in liquid nitrogen and hydrogen, in *Advanced Bearing Technology, NASA SP-38*, 1964, 289.
21. **Hady, W. F., Allen, G. P., and Johnson, R. L.**, Wear and friction of mechanical carbons in liquid oxygen as influenced by transfer films, *ASLE Trans.*, 6, 201, 1963.
22. **Hady, W. F., Allen, G. P., Sliney, H. E., and Johnson, R. L.**, *Friction Wear and Dynamic Seal Studies in Liquid Fluorine and Liquid Oxygen*, NASA Tech. Note D-2453, 1954.
23. **Buckley, D. H. and Johnson, R. L.**, Mechanism of lubrication for solid carbon materials in vacuum to 10^{-9} millimeter of mercury, *ASLE Trans.*, 7, 97, 1964.
24. **Kellogg, L. G.**, Developments in high temperature ultra high-vacuum friction studies, *Lubr. Eng.*, 22, 57, 1966.
25. **Nickerson, D. B.**, U.S. Patent 2,892,663, 1959.
26. **Nickerson, D. B.**, U.S. Patent 2,897,025, 1959.
27. **Vitelozzi, W. J.**, *Report of an Investigation on Water and Steam Lubricated Bearings*, EES Rep. C-3229C, U.S. Naval Engineering Experiment Station, 1963.
28. **Cornell, R. L., Lucas, R. S., and Young, H. L.**, *An Investigation into the Wear and Friction Characteristics of Ship Propeller Shaft and Seal Material Combinations*, M.S. thesis, Massachusetts Institute of Technology, Cambridge, 1960.
29. **Summers, E. R.**, Thrust Bearing, U.S. Patent 3,063,598, 1962.
30. **Scott, M.**, Self-Aligning Bearings, U.S. Patent 3,160,449, 1964.
31. **Shaw, M. C. and Macks, E. F.**, *Analysis and Lubrication of Bearings*, McGraw-Hill, New York, 1949, 240.
32. **Tower, B.**, First report on friction experiments, *Proc. Inst. Mech. Eng. (London)*, 34, 632, 1883.
33. **Reynolds, O.**,*Scientific Papers, Vol. 2, 1881—1900*, Cambridge University Press, London, 1901.
34. **O'Connor, J. J. and Boyd, J.**, *Standard Handbook of Lubrication Engineering*, McGraw-Hill, New York, 1968, 5-50.
35. **Rippel, H. C.**, Cast Bronze Bearing Design Manual, Cast Bronze Bearing Institute, Inc., Cleveland, Ohio, 1967.
36. **Wepfer, W. M. and Cattabiani, E. J.**, Water lubricated bearing development program, *Mech. Eng.*, May, 413, 1955.

37. **Dayton, R. W., Allen, C. M., and VanDyke, H. A.,** *The Performance of Water Lubricated Sleeve Bearings,* Battelle Memorial Institute Report BMI 843, U.S. Atomic Energy Commission, 1953.
38. **Nudelman, H. B. and Sump, G. H.,** *Materials for Water Lubricated Bearings,* No. TID 5187, U.S. Atomic Energy Commission, 1954.
39. **Heller, S., Shapiro, W., and Decker, O.,** A porous hydrostatic gas bearing for use in miniature turbomachinery, *ASLE Trans.,* 14, 144, 1971.
40. **Burr, M. C.,** Controlled Clearance Carbon Bearings, U.S. Patent 2,901,300, 1959.
41. **Brennan, J. S.,** Short Life Bearing, U.S. Patent 2,866,669, 1958.
42. **Siehert, P. G.,** Bearings and Shroud Combination, U.S. Patent 3,022,126, 1962.
43. **H. Kranz Verwaltung-GmbH,** A Carbon Bearing for Revolving Shaft, U.K. Patent 1,025,669, 1966.

Chapter 8

ROTARY PUMPS AND MOTORS

I. INTRODUCTION

Carbon is often used for pump and motor elements other than bearings and seals. Typical applications are

(1) Vanes
(2) Rotors
(3) End plates
(4) Wear shoes
(5) Apex seals

Carbon should be considered for sliding elements wherever the fluid that is being pumped is a poor lubricant. The most common examples are air pumps or blowers. However, certain liquids such as water, gasoline, liquified hydrocarbons, refrigerants, and solvents also have such low viscosities that they cannot form fully effective hydrodynamic load-bearing films at normal rubbing speeds. Formation of a load-carrying hydrodynamic film is also inhibited when the liquid being pumped boils or cavitates near the rubbing surface.

Carbon also merits consideration where the fluid being handled is highly corrosive. Acid pumps, for example, use carbon components in lieu of those made from expensive alloys. Because of the inertness of carbon to all but strongly oxidizing acids, its use in pump applications extends pump life and minimizes contamination of the pumped fluid.

II. VANE PUMPS FOR GASES

Figure 8.1 pictures a rotor and vane combination for a high-performance gas pump used on small aircraft. The pump, in this case, handles incoming air at temperatures ranging from −54 to 74°C (−65 to 165°F) and achieves a compression ratio of 4 to 1. Peak output air temperature is 175°C (350°F) and the efficiency of the pump (adiabatic reversible) averages 65%.

Tip speed on the vanes reaches 12 m/s (2300 ft/min), and the pump is subjected to random vibrations of 6 g magnitude at frequencies to 50 Hz. The typical life of this pump is 1000 hr with 98% passing 500 hr. Tip loading on each vane reaches a maximum of 10 kN/m (60 lb_f/in.) during the compression part of each cycle. This load is made up of three major components: a centrifugal load due to rotation, an inertial load accelerating the vane back into its slot, and a friction vector resisting the movement of the vane back into its slot. Tip load due to gas pressure is negligible because the vane is so thin (less than 3 mm).

Contact of the vane with its housing occurs, theoretically, on a line which shifts across the nose of the vane twice each revolution. The calculated Hertzian stress in this application reaches 550 MN/m^2 (80,000 lb/in.^{2}). To carry loads like this, the inner surface of the pump housing must be hard (35 Rockwell C minimum) and smooth (0.13 to 0.26 μm or 4 to 8 μin. AA). Successful liner materials include electrolyzed, fine-grain cast iron, hard chrome plate, and flame-sprayed aluminum oxide. Hard, abrasion resistant carbons, often with a film-forming additive, are used for vanes in pumps like these.

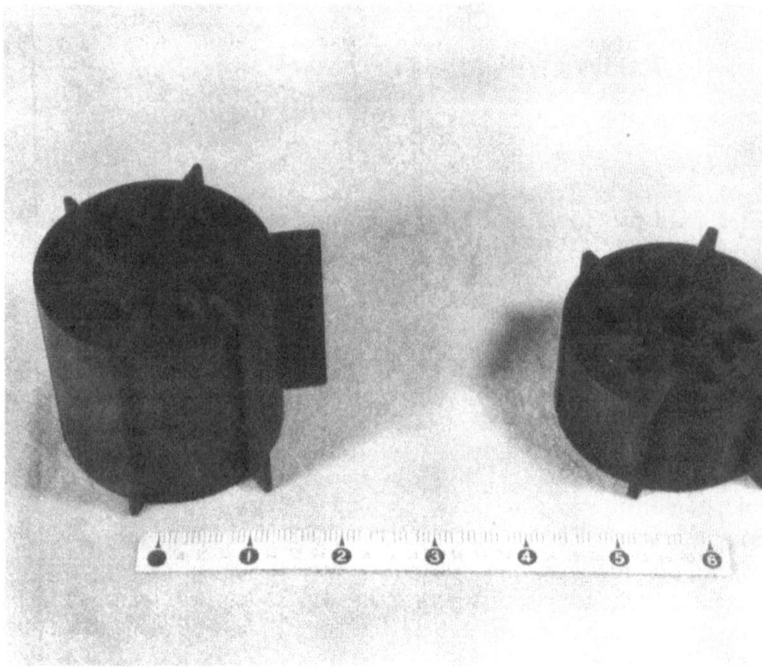

FIGURE 8.1. Carbon rotor and vanes.

III. APEX SEALS FOR ROTARY COMBUSTION ENGINES

Apex seals perform the same function as vanes in the preceding example. Operating conditions, however, are considerably more demanding:

Operating condition	Air pumps	Rotary combustion engines (1)
Gas temperatures, °C (°F)	−54 to 225 (−65 to 440)	−35 to 1170 (−30 to 2100)
Vane temperatures, °C (°F)	25 to 175 (80 to 350)	25 to 425 (80 to 800)
Compression ratio (Maximum)	5:1	10:1
Peak gas pressures, MN/m^2 ($lb/in.^2$)	0.35 (50)	5 (720)
Tip loading, MN/m (lb/in.)	0.01 (60)	0.035 (200)
Peak rubbing speed, m/s (ft/min)	15 (3000)	20 (4000)
Lubricant	None	Trace oil
Life, hr	500 to 1000	500 to 2500

Metal-bonded carbons are presently the most widely used apex seal material with metal-impregnated carbon-graphites as the second most widely used. The most common metals are titanium, aluminum, and copper-based alloys. The present apex seal materials are barely acceptable, and considerable research is underway for an improved material.

IV. LOW PRESSURE, HIGH VOLUME AIR PUMPS

Figure 8.2 shows a different type of air pump. The pump is used to compress secondary air injected into automobile exhaust manifolds and catalytic converters. A relief valve limits discharge pressure to 0.1 MN/m^2 (15 $lb/in.^2$), a pressure that is reached

FIGURE 8.2. Automotive air pump.

at a rotational speed of 600 rpm. Peak internal air temperature is 150°C (300°F), peak speed is 7000 rpm, and power consumption at that speed approaches 11 kW (15 hp).

Figure 8.3 presents a cross section of this pump showing details on one of the vane-vane shoe configurations. The pump was designed so that the rotor is driven eccentric to a stationary shaft and pump housing. Pumps are manufactured with either two or three vanes which are attached to the stationary shaft. The vanes extend through clearance slots in the rotor. Vane tips contact the pump housing, but centrifugal forces of the vane tips against the pump housing are limited by the attachment of the vanes to the stationary shaft. The carbon-graphite vane shoes seal by making contact along the plastic vane sides while the smaller carbon-graphite thrust shoes seal by making contact with the plastic vane ends.

One of each pair of vane shoes drives the rotating vane, and for this reason it is called the "vane driver shoe". It is the most heavily loaded carbon member, and its wear controls the service life of the pump. The other vane shoe is required only to seal, and thus is called the "vane follower shoe". The vane follower shoe is spring-loaded to assure contact at all times with the vane.

Production pumps for the 1975 automobile model year (U.S.) used glass-filled phenolic vanes. Contact force between vane and carbon vane shoe was kept below 1.6 kN/m (9 lb/in.) length, and the movement of carbon along the vane flank was 35 mm (1.36 in.) per revolution. Maximum allowable wear on the most critical member, the

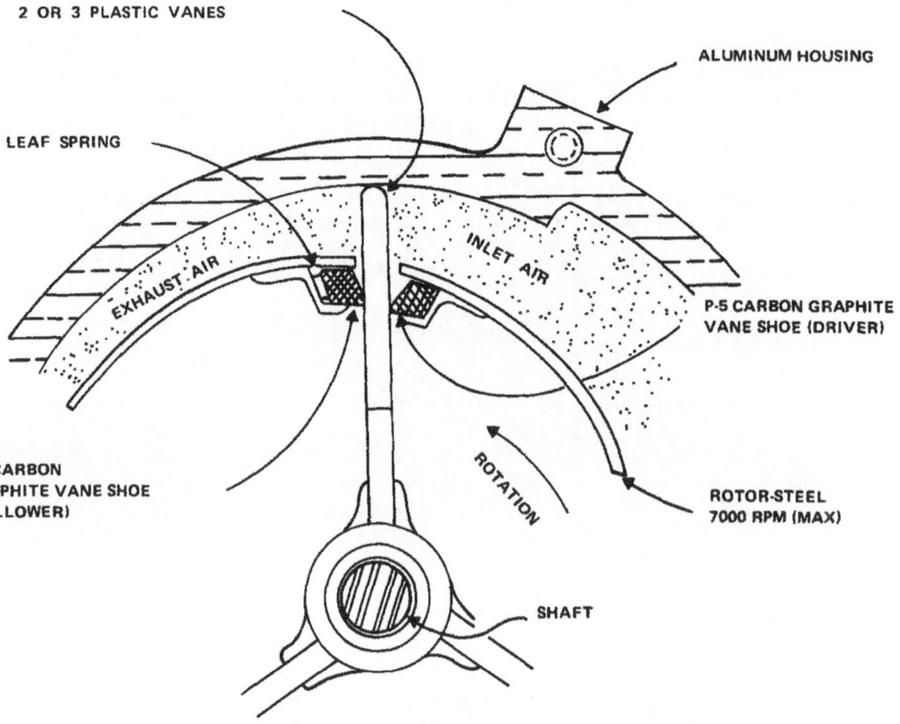

FIGURE 8.3. Vane detail—automotive air pump. (From Shobert, W. R., Carbon-Graphite as a Solid Lubricant Against Plastic, in *Proc. AFML-MRI Conf. Solid Lubricants*, Midwest Research Institute, Kansas City, Mo., September 9 to 11, 1966, 332.)

vane drive shoe, is 0.76 mm (0.030 in.). Service experience with millions of pumps indicates an average life in excess of 100,000 car miles. This is equivalent to a carbon wear of 4.8×10^{-11} m/m rubbed (0.6×10^{-9} in./ft rubbed). It is interesting to note that linear travel between the phenolic vane and the carbon vane shoe is approximately 10% of the distance traveled by the automobile.

V. END CLEARANCE FOR GAS PUMPS

End clearance is critical in controlling internal leakage and, hence, efficiency of vane-type gas pumps. The design problem is aggravated by the fact that metal pump housing usually grows faster than the carbon vanes as the pump heats up. Consider the following situation for a 4-in. vane in a pump where the temperature rises from 20 to 175°C (70 to 350°F) after it has operated 15 min.

Initial (20°C) vane length is 100.000 mm.

Final (175°C) vane length is

$$100.00\,[1 + (175-20)\,5.0 \times 10^{-6}] = 100.000\,(1 + 7.8 \times 10^{-4}) = 100.078 \text{ mm}$$

where 5.0×10^{-6} °C^{-1} is the coefficient of thermal expansion for the carbon blade material.

Initial housing (internal) length is 100.005 mm.

Final housing length is

$$100.005\,[1 + (175-20)\,11 \times 10^{-6}] = 100.178 \text{ mm}$$

where 11×10^{-6} °C^{-1} is the coefficient of thermal expansion for the metal housing. Initial (20°C) clearance, each end, is 0.0025 mm. Final (175°C) clearance, each end, is 0.050 mm. A 20-fold increase in end clearance would typically give an 8000-fold increase in end leakage. The usual way around this dilemma is to use a rotor/vane combination with identical CTEs and spring load one end plate to maintain rubbing contact. In addition, it is desirable to keep the axial length of the pump as short as possible while maximizing volumetric capacity through increased displacement and rotational speed.

VI. FLANK WEAR ON VANES

Flank wear on the low pressure side of each vane as it moves in and out of the rotor slot is another problem requiring careful attention by the designer. To keep blade tilt to a minimum, clearance between the slot and the vane should not exceed 0.025 mm (0.001 in.). The trailing (or low pressure) edge of the slot should be hard and smooth and have a small (0.1 mm or 0.004 in.) radius to avoid shaving the vane. Both slot and blade must be straight and free of warpage under all operating conditions.

VII. USE OF CARBON VANE AIR PUMPS AND MOTORS

Carbon vane pumps have many advantages for pumping compressible fluids, but their design, manufacture, and use requires painstaking attention to details. Blade breakage is always possible, and since this can lock the rotor, the drive should be fitted with a shear pin (or clutch) to protect the rest of the assembly.

Carbon vane pumps are less sensitive to the ingestion of fine solids (dust, etc.) than oil-lubricated metal vane pumps. Still, operation in a dusty environment with an unfiltered intake will definitely shorten pump life.

Small amounts of oil mist are particularly damaging to carbon vane pumps and air motors. Under these circumstances there is not enough to form a load-carrying hydrodynamic film. Instead, the oil combines with carbon wear particles to form a high-friction film on all rubbing surfaces. This increases both vane drag and operating temperatures and usually leads to early failure. There have been many premature failures of carbon-vane air motor hand tools caused by oil laden compressed air. Removal of *all* oil from compressed air is so difficult that one must usually start with a complete new oilfree air supply system to get optimum performance from these types of air motors.

VIII. VANE PUMPS FOR LIQUIDS

When liquids are handled, maximum vane load rather than rising temperatures becomes the main concern. Slower rotational speeds and thicker vanes which project less from the rotor are normal. Because liquid viscosities are two orders of magnitude higher than gas viscosities, greater clearances can be used without concern for internal leakage.

A factor that is often overlooked is the dimensional stability of carbon vanes in different liquids. Many grades of mechanical carbon will swell as much as 0.05% when immersed in polar liquids such as water, methyl alcohol, or dimethylformamide. This swelling is reversible. The tendency to swell can be greatly reduced by heat treatment of the carbons to eliminate their micropore structure and drive out the last trace of bonded hydrogen. This usually requires a firing temperature above 1095°C (2000°F).

IX. CARBON MATERIALS FOR VANES, ROTORS, AND END PLATES

Five factors should be considered in choosing materials for this group of applications.

Flexural strength — Flexural strength affects required vane thickness and amount of permissible projection from the slot. It should be as high as possible.

Density — Density affects centrifugal load on vane tips. It should be as low as possible.

Coefficient of friction — Coefficient of friction controls the drag force on vane tip.

Wear rate — Wear rate controls pump life.

Dimensional stability — Dimensional stability is important for all close tolerance parts.

Of lesser importance, but still warranting consideration, are coefficient of thermal expansion and the thermal conductivity. When one compares the above properties of mechanical carbons to those of most metals, it is obvious why carbon finds such wide use in self-lubricated pumps and motors.

X. NOTCH SENSITIVITY OF CARBON PARTS

Carbon vanes in air pumps are subjected to repeated high loads as the vanes flop from side to side in the rotor slot and move in and out radially. Scratches along the vane flank, together with chips along the edges of vanes, are difficult to avoid during manufacture. Even "perfect" vanes quickly develop scratches and small chips after running a few hours. This leads us to consider the effect of chips and scratches on vane strength.

When the ASTM committee on manufactured carbon, ASTM C-5, was developing test methods for flexural and tensile strength of manufactured carbon, the role of chips and scratches was explored. Surface finishes ranging from highly polished to rough sawed textures were investigated using test specimens from several types of manufactured carbon. The investigators concluded that as long as the finish roughness was no greater than the inherent roughness due to grain size of the filler, no effect of finish upon the strength was discernable. Manufactured carbon differs from glass and many ceramics in this respect.

XI. FATIGUE RESISTANCE OF CARBON

Another feature of air pump designs is the repeated imposition of bending stresses upon the vane. Fatigue effects in metals often reduce the usable stress levels to 30% or less of their initial strength values. Repeated stressing supplies enough energy to cause grain growth and, in some instances, migration of alloying elements to the grain boundaries. This weakens the part.

Carbon, however, has interatomic bonding energies so high that grain growth or migration of crystal defects is virtually impossible to obtain. Accordingly, one would expect manufactured carbon and graphite bodies to have excellent fatigue endurance. A review of recent tests shows this to be the case.

Low-cycle endurance tests were performed on a graphite reed by Green.[3] After 10,000 flexures, the average strength of AUF graphite was 74% of its average starting strength. Green's tests were done on a relatively coarse-grained graphite having an initial flexural strength below 25 MN/m^2. Kachur and Vandergrift[4] tested HLM-85 and H-206-85 graphite thin-walled rings. These were compressed diametrically at 35 cps. After 10,000 cycles, average residual strength ranged from 70 to 85% of the initial value, but again the initial strength of his specimens was relatively low, 25 to 30 MN/

m². Lichter and Robinson[5] ran their tests on AXF graphite out to 10 million cycles and found a 65% retention of 1-cycle strength. Again, the tests were performed on a relatively weak, coarse-grained graphite.

Fine-grained, high-strength graphites were tested for fatigue endurance by Barabanov et al.[6] and Kucher et al.[7] Barabanov used rods 227 × 17 mm with a central gage section 15 mm in diameter. His specimens were machined from extruded graphite having a density of 1.82 to 1.88 g/cc. One set of rods was cut with the axis parallel to the direction of extrusion; the other set was cut with the axis transverse to the extrusion direction. Fatigue strength, after 10 million cycles, ranged from 40 to 50% of the initial value.

Tests of Kachur et al.[7] were made on P-03, a high-density molded graphite with a flexural strength of 55 MN/m². Their tests were conducted at room temperature on a Baldwin-Hamilton model SF-2 constant load machine. Vibration frequency was 1800 cpm, and the bending load was varied sinusoidally. The test specimen was 6 mm thick with an overall length of 100 mm and a maximum width of 50 mm. After the test was completed, flexure bars 6 mm × 12 mm × 50 mm were cut from the end of the specimen and tested with a 12-mm inner span and 37-mm outer span to obtain the initial strength value. Figure 8.4 shows the results of both the tests of Barabanov et al. and of Kachur et al.

Both Barabanov and Kachur report that many of their specimens broke while starting the test machine. Kachur, for example, reports that a specimen was unbroken after 10^6 cycles at 36 MN/m², but it broke during the start-up when it was retested at this same load. This suggests that transients during start-up may generate loads in the specimen that are higher than calculated. Graphite has a strain to failure value of only 0.1

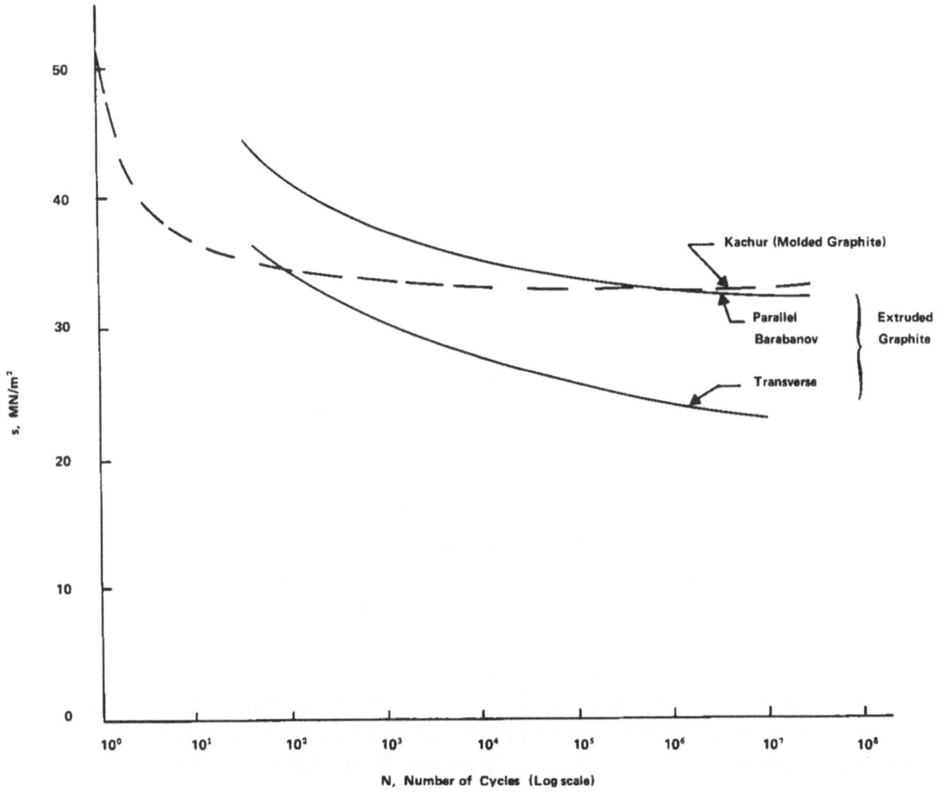

FIGURE 8.4. Fatigue strength of graphite.

to 0.3%; thus, onetime transients that do not affect a metal specimen will overload and rupture graphite.

The terminal fatigue strength of most manufactured graphites is thought to be between the 45% reported by Barabanov and the 85% reported by Kachur and Vandergrift. Further, it seems reasonable that low-strength, low-modulus graphites should retain a higher fraction of their initial strength after repetitive stressing. Such graphites already contain a multitude of microcracks and flaws capable of absorbing energy. This attribute may be the reason for the success of coarse-grained graphite in rupture discs subjected to repeated pressure cycles.

Designs which expose manufactured carbon parts to repeated high loads must consider not only fatigue effects but also piece to piece variations in initial strength. Wilkins and Reich[8] applied Weibull statistical methods to this problem. They concluded that proof testing at a stress level high enough to weed out the low breakers, say 1 to 10% of the population, greatly increased the reliability of the surviving population. This agrees with the author's observation that if a carbon vane survives the first minute of operation, it will wear out before it breaks.

XII. IMPACT ENERGY ABSORPTION

Manufactured carbon has a low impact strength. It does not flow or plastically deform under stress at temperatures below 2200°C. The tensile stress-strain "curve" at room temperature is essentially linear, terminating at the point of rupture. Thus, the energy absorbed at rupture in tension can be readily calculated:

$$M_t = P_t^2/2E$$

where M_t = energy absorbed at rupture, J/cm^2; P_t = ultimate tensile strength, MN/m^2; and E = modulus of elasticity, MN/m^2.

Taking carbon grade P-29 (see Table 4.2) as an example,

$$M_t = 55^2/2\ (23{,}000) = 0.066\ \text{J/cm}^2 = 0.8\ \text{ft lb/in.}^2$$

Charpy and Izod tests break the specimen in flexure, not tension. Table 8.1 shows that the energy absorption during a "slow" break is not much different from that of a "fast" break. This is in sharp contrast to what one observes for most metals and plastics.

The "slow break" specimens of Table 8.1 came from different production lots than the "fast break" specimens; hence, one should be careful about grade by grade comparisons. All specimens were within current production tolerance for such properties as density, hardness, and electrical resistivity, but the "slow break" P-5 blocks were at the low end of these limits.

Taking all factors into consideration, I believe reducing the load-to-break time from 30 to 0.0002 sec results in no more than a 30% reduction in required break force for fine-grain mechanical carbons like these. Specimen temperature probably has little effect on this reduction since the mobility of crystal boundaries is very low below 2000°C.

All of the energy absorption values in this series of 24 tests were between 0.05 and 0.23 J/cm^2 of break surface (0.24 and 1.09 ft lb/in.^{2}). Basically, the relatively low energy absorption capabilities of carbon derive from its low strain to failure characteristic. Designs which recognize this have little to fear from shock loading.

TABLE 8.1

Energy to Break Carbon Blocks[a] in Flexure

Grade	P-03	P-5	P-9
Slow break			
Support span = 38 mm; head speed = 8.5×10^{-6} m/sec			
Number tested	5	5	5
Average deflection at break ($m \times 10^{-6}$)	278	262	281
Standard deviation ($m \times 10^{-6}$)	35	38	40
Average force at break (N)	1983	1909	2280
Standard deviation (N)	185	195	74
Average energy to break (J)	0.277	0.230	0.320
Standard deviation (J)	0.056	0.077	0.052
Fast break[b]			
Support span = 40 mm; head speed = 1.03 m/sec			
Number tested	2	3	4
Average deflection at break ($m \times 10^{-6}$)	205	194	182
Standard deviation ($m \times 10^{-6}$)	57	21	11
Average force at break (N)	1076	2063	1819
Standard deviation (N)	449	236	240
Average energy to break (J)	0.176	0.253	0.217
Standard deviation (J)	0.115	0.039	0.043

[a] Block size = $12.7 \times 12.7 \times 50.8$ mm.
[b] Data obtained by ETI Products, Inc., Santa Barbara, Calif. 93111.

REFERENCES

1. **Norbye, J. P.**, *The Wankel Engine*, Chilton Book Company, Philadelphia, 1971.
2. **Shobert, W. R.**, Carbon-Graphite as a Solid Lubricant Against Plastic, in Proc. AFML-MRI Conf. Solid Lubricants, Midwest Research Institute, Kansas City, Mo., September 9 to 11, 1966, 332.
3. **Green, L., Jr.**, The failure of graphite under alternating stresses, *ASME, J. Appl. Mech.*, p. 345, December 1951.
4. **Kachur, V. and Vandergrift, E. F.**, Low Cycle Fatigue in Structural Graphite Under Localized Loading, WANL-TME-1764, March 1968.
5. **Lichter, H. and Robinson, E.**, Fatigue Behavior of EP-1924 Graphite and a Generalized Design Criterion, UCRL-50337, University of California, Berkeley, 1967.
6. **Barabanov, V. N., Anufriev, Y. P., Zaitsev, G. G., and Pimkin, M. Y.**, Some special features of the method of testing graphite for fatigue in alternate-reversing bending and the results obtained, *Zavod. Lab.*, 32(4), 459, 1966.
7. **Kachur, V., Layland, E. L., and Lorenz, F. R.**, Properties of P-03 Graphite, WANL TME 1031, Unclassified NERVA Res. Dev. Rep., 1967.
8. **Wilkins, B. J. S. and Reich, A. R.**, An Estimation of the Probability of Fatigue Failure of Some Graphites, AECL-3958, Whiteshell Nuclear Research Establishment, Pinawa, Manitoba, Canada, February 1972.

Chapter 9
METERING DEVICES

I. INTRODUCTION

Manufactured carbon is often used in metering devices. Typical carbon components are

(1) Bearings
(2) Rotors and vanes
(3) Pistons
(4) Valve seats

Gasoline, chemical solvents, liquified gases (including ammonia), strong acids, cola syrups, and human blood are but a few examples of fluids which commonly require carbon parts in the meter.[1]

II. METERS

Figure 9.1 shows carbon pistons used in chemical metering devices. The pump using the smaller piston is capable of metering acetone, naphtha, phenol, or liquid ammonia with an accuracy of ±1%. The cylinder in which the piston rides is either bronze or 316 stainless steel, depending on the chemicals to be handled. The finish on the inner surface of the cylinder must be better than 0.2 to 0.3 μ (8 to 12 μin.) rms. Other operating parameters are

Temperature	−54 to 71°C (−65 to 160°F)
Pressure differential	40 KN/m^2 (6 lb/in.^{2})
Strokes	3.3Hz (200 cycles/min)
Stroke length	76 mm (3 in.)
Piston diameter	76.200 ± 0.005 mm (3.0000 ± 0.0002 in.)
Displacement	4587 cm^3/sec (72 gal/min)
Average rubbing velocity	0.50 m/sec (1.6 ft/sec)
Minimum life	37,800 m^3 (10 million gal)

Another type of liquid meter uses a rotating cavity instead of a reciprocating piston. In this type of meter, a leakfree seal must be made between the rotating element and the stationary parts of the meter. A carbon part is often used as one of the rubbing members because carbon is self-lubricating, dimensionally stable, and wear resistant.

Figure 9.2 shows two types of valve seats used in gasoline-dispensing pumps. The one on the left rotates while the one on the right reciprocates. The metering cavity in the reciprocating seat, not visible in this illustration, has a capacity of approximately 60 cm^3. The mating face for these valve seats is either 17-4ph stainless steel or fine-grained cast iron in order to combat the corrosive effect of entrained water often present in gasoline. Surface finish is 0.2 to 0.4 μm (8 to 16 μin.) rms on the metal. Other operating parameters are

Temperature	−54 to 49°C(−65 to 120°F)
Fluids metered	All types of automobile fuel
Pressure differential	17 KN/m^2 (2.5 lb/in.^{2})
Contact (squeezing pressure)	30 to 100 KN/m^2 (4.5 to 15 lb/in.^{2})
Rubbing speed (average)	3 to 4 m/sec (10 to 15 ft/sec)
Life required	20 yr or 37,800 m^3 metered (10 million gal)
Metering accuracy	± 1% at 10^{-3} to 10^{-4} m^3/sec (1.5 to 15 gal/min)

FIGURE 9.1. Carbon pistons for chemical metering pumps.

Metering devices handling foodstuffs generally operate at lower speeds and higher contact pressures than those in gasoline pumps. A further requirement is the need for low porosity in the carbon member in order to avoid sites for growth of mold or bacteria.

Figure 9.3 pictures a metering valve from a coin-operated drink dispenser. Valves like this one are used to meter a wide variety of potable liquids such as cola syrups, carbonated water, coffee, milk, and alcoholic beverages. Similar valves are used in blood pumps and other life support devices. In all of these cases, carbon is selected because of its chemical inertness, its lack of toxicity, its dimensional stability, and its low wear rate.

III. ORIFICES AND NOZZLES

A special case of metering is the use of orifices or nozzles for throttling the flow of extremely corrosive or high-temperature fluids. Rocket nozzles, the most important application of this type, fall outside the scope of this book. However, carbon or carbide-coated carbon can also be used as liners for blow down nozzles on boilers and high-pressure reactors. Carbon-based materials are available which are as hard as cobalt-bonded tungsten carbide, yet are far more resistant to chemical attack or thermal shock.

IV. FLOW-INDICATING DEVICES

There are many industrial applications where metering is not required; instead, all that one needs is an indication of the direction and approximate velocity of flow.

FIGURE 9.2. Carbon metering valves for gasoline pumps.

Figure 9.4 pictures flappers and rotors used in flow-indicating devices of this type. Carbon has proven ideal for these applications since it can be "molded to size", is unaffected by most chemicals, rotates freely without lubrication, and has extremely low wear rates.

V. SELECTION OF GRADES

Dimensional stability, resistance to chemical corrosion, and moldability are the three most important considerations in selecting carbon grades for any metering device. Clearances are generally small to minimize leakage through the meter, but pistons must not swell and stick in the cylinder; valve seats must not warp because any distortion can cause binding or leakage.

Most meters must be capable of handling a wide range of fluids. The inherent chemical inertness of carbon is a great advantage in satisfying this requirement.

The choice of a carbon grade that can be molded to size is important for parts having irregularly shaped cavities or ports. Volumetric tolerance of ±1% at a cost well below that of machining can be realized on parts produced in quantities greater than 500 per year.

Factors of minor importance in grade selection are the permeability, porosity, friction coefficient, and wear rate of the carbon. Low permeability in carbon parts facilitates the generation of a hydrodynamic film and leads to lower friction and wear rates on meters with high rubbing speeds. Low porosity is desirable where beverages or body fluids (like blood) are handled, since sterilization between uses is less difficult when there are few internal recesses and pores in which microorganisms can lodge.

FIGURE 9.3. Carbon metering valve parts from a drink dispenser.

The friction coefficient is important only in small meters handling gases or low viscosity fluids.[2] Usually there is ample pressure gradient available to drive the meter and overcome sliding friction at the metering interface.

VI. FAILURE ANALYSIS

Meters are precision instruments and should be treated with the care and respect due any instrument. A meter which is not performing properly should not be taken apart; instead, it should be returned to the manufacturer for disassembly and repair.

Failures involving carbon parts usually occur because of broken or swollen carbons. If disassembly of the meter shows the carbon parts are broken, the metering chamber should be carefully inspected for small fragments. The absence of fragments, particularly when worn or shiny areas along the break are also observable, suggests that the meter ran some time after the break occurred. Cause of the breakage is often difficult to determine in these cases. Nevertheless, the carbons should be removed and checked for deep scratches or imbedded foreign particles on the rubbing surfaces. Ingestion of solids into a liquid meter can lead to loss of running clearances, thereby jamming the meter and breaking the carbon parts.

The carbons taken from the meter should also be examined for evidence of high contact pressures at drive points such as ears or notches. Wear or chipping at these sites suggests overloading caused by loss of clearance between moving and stationary parts.

If the carbon part is intact, it should be removed and carefully measured. If it is oversize or even if it conforms to print dimensions, it could be a "sweller". The tendency to swell can be checked by measuring the carbon dry, soaking it in the liquid to

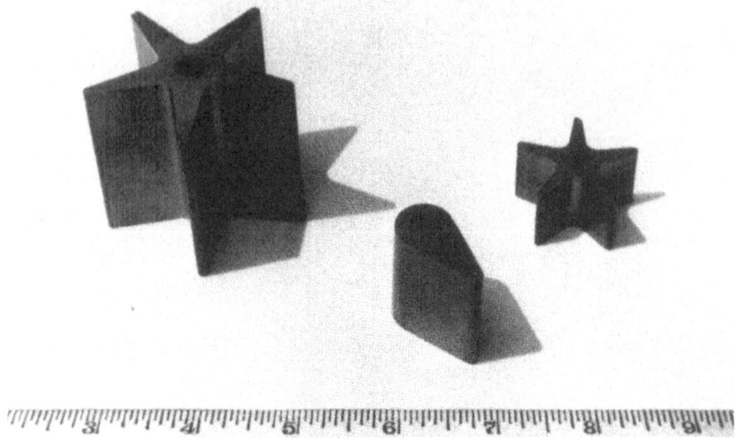

FIGURE 9.4. Flow-indicating rotor and flap.

be metered for at least 24 hr, and remeasuring it while still wet. Any increase in size indicates the carbon is dimensionally unstable in the liquid used for testing. At this point the carbon vendor should be contacted for assistance in selecting a more stable grade of carbon.

REFERENCES

1. **Lyddon, P. E. and Hurden, R. K.,** Some mechanical engineering applications of carbon, in *Industrial Carbon and Graphite,* Society of Chemical Industry, London, 1958, 579.
2. **Wisander, D. W. and Johnson, R. L.,** Wear and Friction in Liquid Nitrogen and Hydrogen of the Materials Combinations Carbon-Stainless Steel and Carbon-Carbon, in Proc. of the AFNI Propulsion Systems Lubrication Conf., 1961, 157.

Chapter 10

PISTON RINGS AND CYLINDER LINERS

I. INTRODUCTION

One of the earliest uses of manufactured carbon in mechanical devices was as piston rings for nonlubricated compressors. For many years manufactured carbon was the only material used in this type of compressor. Since 1960 there has been growing use of carbon-filled PTFE rings for low-pressure, low-temperature applications. For temperatures above 200°C(400°F) or pressures above 15 MN/m^2 (2177 psi), however, carbon rings are still without peer. Examples of high-temperature, high-pressure applications can be found in high-performance compressors and actuators for aircraft flight control surfaces.

Carbon remains the preferred material in applications where the fluids or the surrounding environment are loaded with abrasive dust or grit. The seal ring on an aircraft hydraulic actuator where the extended rod is exposed to grit illustrates this type of application. Carbon also makes an excellent scraper ring for these applications because it can remove adhering grit without affecting the surface finish of the rod itself.

Figure 10.1 shows a typical scraper ring design which uses a garter spring to maintain contact with the shaft. (The part shown is *not* assembled correctly; the double scratches on the chamfer should be assembled side by side.) During each deployment, this ring is exposed to dust laden air mixed with jet exhaust gases at 480°C(900°F) for 60 sec. Rod travel is normally 0.3 m (12 in.) each way during deployment. Ring life, based on laboratory tests, is estimated to be in excess of 10,000 cycles.

II. PISTON RINGS

Figure 10.2 shows typical piston ring components. The thicker one is the piston wear ring (or bull ring) which carries the sliding weight of a horizontal piston. The smaller part is one of the three segments used for the piston rings. Step joints, lapped to give "light tight" contact, can be supplied for applications where leakage must be minimized. Such sets have marked joints to show proper assembly because no two joints are axactly alike. Joints between sections may be "butt," "scarf," or "step" configuration.

The parts shown in Figure 10.2 are designed for a nonlubricated air compressor operating at 0.76 MN/m^2 (110 lb/in.^{2}) discharge pressure and having an average sliding speed of 3 m/sec (600 ft/min). Spring steel expanders add an estimated 0.14 MN/m^2 (20 lb/in.^{2}) to the radial contact pressure between carbon and cylinder liner. Typical wear rates for a premium carbon grade in such applications would be approximately 0.5 mm/1000 hr (0.020 in./1000 hr) or 4.6×10^{-11} m worn per metre rubbed.

III. SELF-EXPANDING PISTON RINGS

The segmental piston ring with its steel expanders is a relatively expensive assembly of parts and requires painstaking care to properly match segments during installation. (Figure 10.2b has match marks — not visible in the photograph — which must be placed together during assembly; furthermore, segments from different ring sets must not be mixed.) A second drawback is the hazard presented by the expander ring. If the carbon ring breaks up due to ingestion of a slug of water, for example, the hardened steel expander ring can score the cylinder liner.

FIGURE 10.1. Carbon scraper ring for aircraft hydraulic actuator.

a) b)

FIGURE 10.2. Piston rings. (a) Support or bullring. (b) Piston ring segment.

One-piece self-expanding piston rings have been developed to avoid these problems. Figure 10.3 shows a typical example. One-piece rings are carefully opened, slipped over the piston head, dropped into the ring groove, and compressed as the piston is installed in the cylinder. The simple appearance of these one-piece rings disguises a sophisticated design procedure.

IV. DESIGN OF SELF-EXPANDING PISTON RINGS

In a typical application, one wants the ring to be as thick as possible in the radial direction, to exert as much outward sealing pressure on the cylinder wall as possible, to be round when installed, to have a minimum gap, and of course, to slip over the piston and drop into the groove without breakage. The key physical property is strain to failure in flexure.

During expansion when the ring is slid over the piston, the innermost fibers are placed in tension. During compression when the piston with its installed ring is slid into the cylinder, the outermost fibers are placed in tension. This latter stress is generally controlled by the added requirement of minimum operating gap width, and this in turn usually means the ends are butted together during installation. Equations gov-

FIGURE 10.3. Self-expanding piston ring.

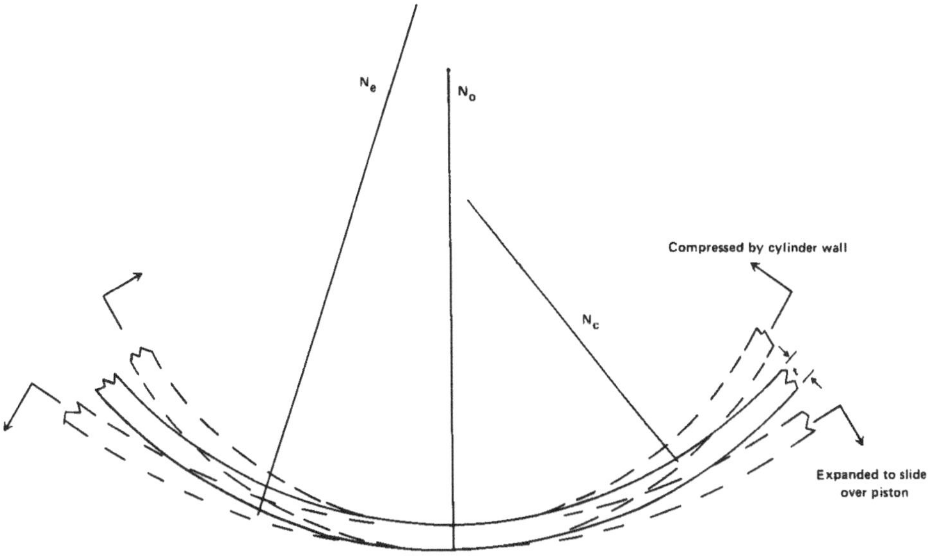

FIGURE 10.4. Self-expanding piston ring radii in normal (N_o), expanded (N_e), and compressed (N_c) states.

erning the forces and strains are given in Timoshenko and MacCullough.[1] Figure 10.4 diagrams these forces.

To idenitfy the parameters which control maximum fiber strain, we must examine an element of the ring.

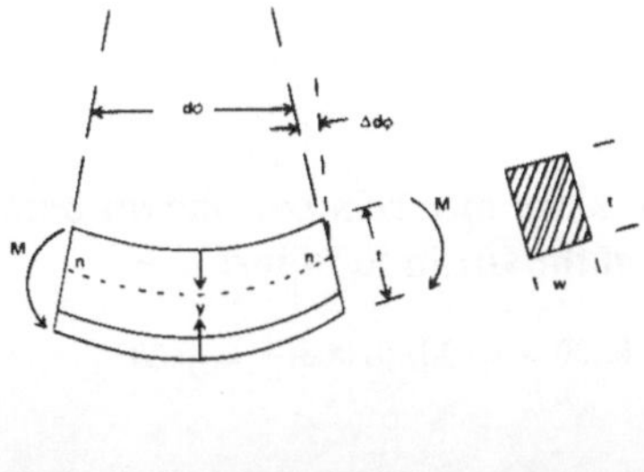

If n-n represents the neutral axis, compression of any fiber at a distance y from this surface is y(Δdϕ), and the corresponding unit compression (dimensionless) is

$$C = \frac{y\,(\Delta d\phi)}{(r - y)\,d\phi}$$

where r = the radius of the neutral surface.

Most piston rings will be rectangular in cross section having a radial thickness t and a width w. In most cases, too, the radius r will also be large compared to the thickness t.

For carbon, elongation of either surface is critical since strain to failure in tension is usually below 1%. Substituting t/2 for y and solving for maximum fiber strain, e max. (dimensionless).

$$e\ max. = \frac{t/2\ (\Delta\, d\,\phi)}{(r - t/2)\,d\phi} \cong \frac{t\,\Delta\, d\phi}{2rd\,\phi}$$

because r is generally very large compared to t/2.

The term Δdϕ will, in practice, be governed by the free gap width. To get maximum radial sealing force, the free gap will be set as large as possible consistent with the need that the gap be small when the ring is installed in the cylinder. Minimum diameter will, of course, be when the ring ends are butted, a condition which gives maximum fiber strain on the outer surface of the ring.

Figure 10.5 shows the relation between these variables in the region of interest. Lines for maximum fiber strain values of 0.1%, 0.2%, 0.4% and 0.8% are shown.

A. Specimen Design Calculation

Consider a case with the following fixed parameters:

1. Cylinder diameter, D_c, $75.000\ ^{+0.025}_{-0.000}$
2. Piston diameter, D_P, $74.950\ ^{+0.000}_{-0.025}$
3. Installation clearance with ring ends butted, mm minimum, 0.025
4. Piston ring material, strain to failure in tension, e_f, 0.0036 mm/mm

To be calculated:

1. Installed gap, mm
2. Free gap, mm
3. Ring radial thickness, mm
4. Maximum usable wear, mm
5. Piston groove depth, mm

Let the ring ends be butted during installation with a ring OD of 74.975 mm to give a diametrical installation clearance of 0.025 mm at minimum cylinder diameter of 75.000 mm.

$$\text{Installed gap} = \pi\,(D_C - D_R) = \pi\,(75.00\ ^{+\,0.025}_{-\,0.000}) - 74.975 = 0.075 \text{ to } 0.150 \text{ mm}$$

Free gap and radial thickness are interrelated. Shown below are a family of values based on a 30% safety factor on the strain to failure, i.e.,

$$e\ max = (1.00 - 0.30)\,(0.0036) = 0.0025\ M/M$$

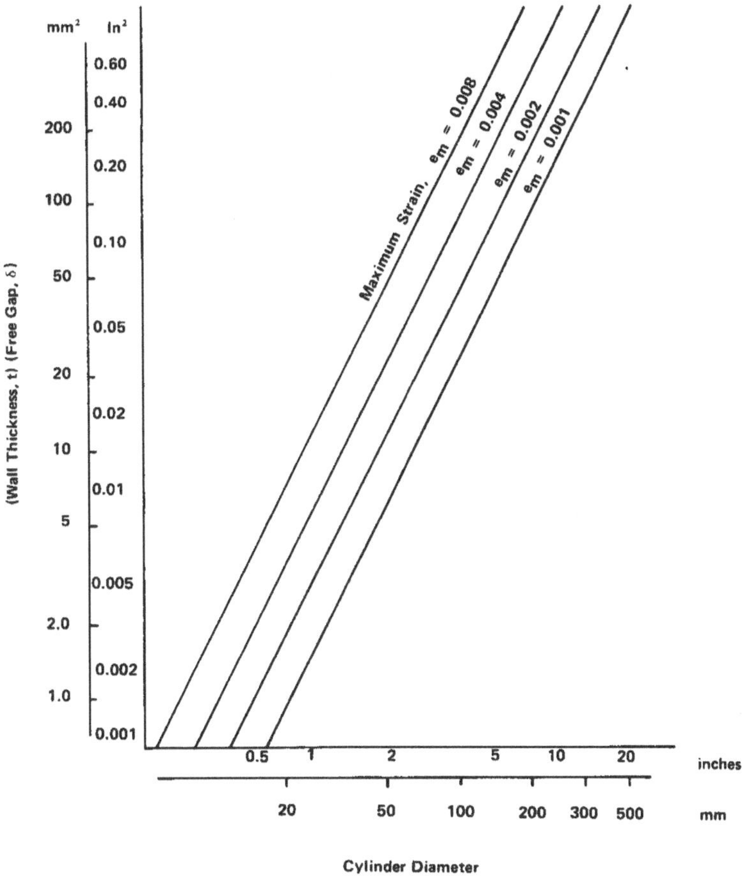

FIGURE 10.5. Self-expanding piston ring design wall thickness.

From Figure 10.5 the product of wall thickness and free gap is 35 mm².

Thickness t (mm)	Gap δ (mm)	Free OD (mm)	Maximum ID (mm)	Minimum groove depth (mm)	Maximum radial wear (mm)
1.0	35+	86.12	95.26	0.98	0.70
2.0	17.5	80.55	81.12	1.98	1.70
4.0	8.75	77.76	72.54*	3.98	3.70

* Less than piston OD requires "built up" piston

$$\text{Free OD} = (\ \delta/\pi + DR = \delta/(\pi) + 74.975 \text{ mm}$$

$$\text{Max ID} = (\text{Free OD} - 2t) + (\delta)/\pi$$

This presumes the gap can be opened as far as it can be closed with equal effect on peak fiber stress.

Minimum piston ring groove depth = t − (minimum piston to cylinder clearance)/2 = t − 0.050/2 mm. Maximum radial wear (assume ring thickness remaining in gap must be at least three times the maximum initial piston to cylinder clearance):

$$\text{Max radial wear} = t - 3(D_c, \text{max.} - D_p, \text{min.}) = t - 3(75.025 - 74.925)$$
$$= t - 0.30 \text{ mm}$$

Using a typical carbon piston ring wear rate of 5×10^{-11} mm wear per millimeter rubbed, even the 1-mm thick ring should travel 1.4×10^{10} mm before reaching minimum thickness. Assuming a 75-mm stroke at 300 strokes per minute, ring life should be over 5000 operating hours.

V. CYLINDER LINERS

Another application for carbon in compressors is cylinder liners.[2] The liner is shrink fitted in a metal housing using techniques described in Chapter 7 of this book. Alternately, the liner can be pressed or cemented into place.

Solid metal pistons are generally used with carbon-lined cylinders. These pistons should be hardened to a minimum of 35 on the Rockwell C scale and finished to 0.2 to 0.3 μm (8 to 12 μin. AA). On small compressors, this type of construction is generally cheaper since it eliminates the cost of piston rings. This construction also finds wide use on fuel and hydraulic pumps.

VI. CLOSING REMARKS

This and the preceding chapters describe a wide range of applications for manufactured carbon in mechanical devices. Many additional applications remain to be discovered, and many improvements in the properties of manufactured carbon remain to be developed. At this time, however, the greatest barrier to more widespread use of manufactured carbon is an information barrier. Most practicing engineers have never heard of manufactured carbon. If this book has helped to break down that one barrier, it has been a success.

REFERENCES

1. **Timoshenko, S. and MacCullough, G. H.,** *Elements of Strength of Materials,* 2nd ed., D. Van Nostrand, New York, 1940, 234.
2. **Peltz, S. J. and Cigal, F. J.,** Wear Resistant Bushings or Liner for the Plunger Cylinder or Bores of Pumps Handling Water — Contaminated Fuel, U.S. Patent 2,733,968, 1956.

Appendix A

GLOSSARY OF TERMS FOR MANUFACTURED CARBON

Ash — "The residual product following oxidation of the base carbon as determined by prescribed methods."*

Binder — "A substance, usually an organic material such as coal tar pitch or petroleum pitch used to bond the coke or other filler material prior to baking."*

Carbon — "A nonmetallic element, atomic number 6, symbol C, molecular weight 12.01115 which exists in several allotropic forms."* The word carbon is also used as a broad term which includes all forms of manufactured carbon and graphite.

Carbon-graphite — A manufactured product which contains some graphite, natural or synthetic, and which has a rigid, hard structure produced by baking at temperatures between 625 and 2225°C (1160 and 4040°F).

Char — A carbonaceous solid produced from cellulose, coal, resins, or other materials by thermal decomposition.

Coke — "A carbonaceous solid produced from coal, petroleum or other materials by thermal decomposition with passage through a plastic state."*

Compressive strength — The ultimate stress, based on cross sectional area, required to rupture the material when subjected to a compressive load.

Electrographite — Any manufactured carbon product which has been heat-treated at a temperature above 2525°C (4580°F).

Extruded graphite — A carbon body which has been shaped by forcing a blend of the binder and carbon filler material through a die under pressure and subsequently heated to a temperature greater than 2525°C (4580°F) to graphitize the structure.

Filler — The ingredients of the carbon mix which retain their particulate identity throughout the processing steps.

First bake carbon — Similar to "carbon-graphite" except that graphite may not be present in the body.

Flaw — A structural defect larger than typical in a carbon or graphite body.

Flexural strength — Calculated ultimate stress required to rupture the material subjected to a bending load.

Grade — "The designation given a material by a manufacturer such that it is always reproduced to the same specifications established by the manufacturer."*

Graphite — A general term which refers to carbon which is crystalline as determined by X-ray diffraction. Crystallite size may range from 10Å upward; and the spacing between layer planes may be as large as 3.45Å.

Graphite (artificial or synthetic) — Same as electrographite.

Green carbon — "A formed, but unfired carbon body."*

Hardness — "The resistance of a material to deformation, particularly permanent deformation, indentation or scratching."*

Impregnated carbon (or graphite) — A carbon, carbon-graphite, or graphite whose pores have been partially filled with an impregnant.

Impregnation — "Partial filling of the open pore structure with another material."*

Lamination — A line of demarcation or crack generally parallel to the principal formed surfaces of a carbon or graphite body.

Machinability — "A measure of the ease with which a material can be shaped with the aid of cutting or abrasive tools."*

Metal-graphite — A composite body made from metal and graphite powder with or without a small amount of supplementary binder.

* From C-709-74a Standard Definitions of Terms Relating to Manufactured Carbon and Graphite, American Society for Testing and Materials, Philadelphia, 1975.

Metallized carbon (or graphite) — An impregnated carbon, carbon-graphite, or graphite whose pores have been partially filled with metal.

Molded graphite — A carbon body which has been shaped by forming a blend of the binder and carbon filler material in a closed die by applying external pressure and subsequently heating said body to a temperature greater than 2800 K (4580°F) to graphitize it.

Natural graphite — Mined graphite such as that produced in Texas, Mexico, Austria, Ceylon, Madagascar, etc.

Oxidation of carbon — "The chemical combination of carbon with oxygen or oxygen-containing compounds."*

Particle sizing — "Segregation of granular material into specified particle size ranges."*

Penetration — "The depths to which one material extends into or penetrates another."*

Permeability — "A property measured by the rate of passage of a fluid under a pressure gradient through a material."*

Porosity — "The percentage of the total volume of a material occupied by both open and closed pores."*

Pyrocarbon (or pryographite) — A carbonaceous solid produced by the thermal decomposition of a carbon-containing gas.

Pyrolytic carbon (or graphite) — A pyrocarbon produced from the thermal decomposition of a carbon-containing gas.

Reactivity — "The rate at which another material will form compounds with carbon or graphite."*

Resin-bonded carbon (graphite) — A manufactured carbon in which the binder is a resin.

Surface finish — "The geometric irregularities in the surface of a solid material. Measurement of the surface finish of carbon and graphite shall not include inherent structural irregularities unless these are the characteristics being measured."*

Synthetic graphite — see Electrographite.

Tensile strength — The ultimate stress, based on the original cross section, required to rupture the material subjected to a tensile (axial) loading.

Void — "An unfilled space enclosed within an apparently solid carbon or graphite body."*

* From C-709-74a Standard Definitions of Terms Relating to Manufactured Carbon and Graphite, American Society for Testing and Materials, Philadelphia, 1975.

Appendix B

TEST METHODS FOR MANUFACTURED CARBON AND GRAPHITE

I. APPARENT DENSITY

Three test methods are used, the selection being dependent upon the size and shape of the part and the accuracy required.

ASTM C-559 — This method requires a rectangular parallelepiped or right circular cylinder large enough so its dimensions and weight can be determined with an accuracy of 0.05%. Generally the smallest dimension must be at least 5.0 mm (0.196 in.).

ANSI C-64.1—1970, par. 4.8 — This method requires shapes as above but places no restrictions on size except that weight must be determined with an accuracy of at least 0.5%. Physical dimensions (in inches) are to be measured to three decimal places. The use of very small specimens can result in understating density because of filler particles being plucked out of the surface during sample preparation. (ASTM method C-559 recognizes this potential error by requiring that the smallest dimension be at least five times the length of the largest particle visible on the specimen surface.)

ANSI C-64.1—1970, par. 4.9 — This is the only method which allows use of an irregularly shaped part such as a seal nose or piston ring. The volume of the part is determined by weighing in air, then in water. Care must be taken to avoid adhering gas bubbles on one hand and penetration of water into the part on the other. The latter is the usual problem and commonly results in overstating the density.

All three test methods require that the part be dried before weighing. This step is often overlooked with the result that density is overstated.

II. ASH

ASTM C-561-69 — uses a ground sample (through 60 mesh) and progressive heating from 500 to 750°C (1382°F) in flowing air over a period of 2 hr. It is held at 750°C (1382°F) for 1 hr, then heated to 950°C (1742°F) for 1 hr. It is given repeated 1/2-hr soaks at 950°C until the ash reaches constant weight to within ±0.0002 g.

ANSI C-64.1—1970, par., 4.14 — requires a nitric acid digestion of metal-containing materials prior to ashing. Metals (and other ash constituents) soluble in 6 normal nitric acid (HNO_3) are removed from the sample before ashing. Ashing then is carried out over a 2-hr period with a final (peak) temperature of 750°C (1382°F). The ash specimen is repeatedly returned to the 750°C furnace until constant weight is obtained.

Obviously ash content by the ASTM method will be much higher for metal-containing carbon and graphite materials.

III. CHEMICAL ANALYSIS

ASTM C-560 — describes the analytical methods to be used on graphite to determine:

Element	Range (ppm)	Sensitivity limit	Reproducibility[a]
Silicon	0.2—10	0.001—0.2 mg/100 m*l*	±4
Iron	0.1—6.0	0.04—0.28 mg/100 m*l*	±5
Calcium	5—60	0.5 ppm	±5
Aluminum	To 1.7	0.02 ppm	±0.1
Titanium	2.5—50	0.2 mg/100 m*l*	±2
Vandium	0.2—2.5	0.1 ppm	±5
Boron	0.06—0.3	0.02 ppm	±20

[a] Percent $\sigma/\bar{x}$.

IV. COEFFICIENT OF THERMAL EXPANSION

ASTM E-228 — Describes the method of determining coefficient of thermal expansion by a quartz tube dilatometer. Many carbons oxidize at a significant rate above 350°C (662°F); hence, it is usually necessary to continuously purge the dilatometer with an inert gas when measuring expansion at higher temperatures. Failure to do this tends to give low CTE values owing to loss of carbon at the dilatometer contacting surfaces.

V. COMPRESSIVE STRENGTH

ASTM C-695 — A right circular cylinder is crushed by axial loading. Length to diameter ratio must be between 1.9 and 2.1 with the diameter being at least 10 times theilength of the largest visible grain. Faces of the cylinder must be flat and parallel within 15 min of arc, and teflon cushion discs 0.05 to 0.25 mm thick are recommended to minimize errors due to friction holding back diametrical strain. Load rate must be such that it takes approximately 30 sec to rupture the specimen.

VI. DENSITY

See Section I, Apparent Density.

VII. ELECTRICAL RESISTIVITY

ASTM C-611 — Uses a strip, rod, tube, or bar whose cross section normal to current flow is uniform within 0.75%, and whose length to width (diameter) ratio is at least 6 to 1. The procedure is designed to give an accuracy of ±2% and, in practice, requires specimens longer than 2 in.

ANSI C-64.1—1970, par. 4.4 — Uses a millivolt meter rather than a null balance method for measuring current. It is suitable for small specimens at some sacrifice in accuracy.

ANSI C-64.1—1970, par. 4.5 — Uses a Kelvin bridge for measuring resistance.

The second of these three methods is the most commonly used since it is rarely necessary to know electrical resistivity closer than ±10%.

VIII. FLEXURAL (TRANSVERSE) STRENGTH

ASTM C-651 — A rectangular parallelapiped with a length to thickness ratio of at least 8 is used. Four-point loading is used with the centrally located load span at least two times sample thickness and support span at least three times the load span. Load rate should be 1.27 mm (0.05 in.) per minute, and it normally takes more than 5 sec to reach the rupture load. Strength — or, more precisely, maximum fiber stress — is calculated using the standard beam formula. Where the load span is one third of the support span this is

$$S = PL/bd^2$$

where S = flexural strength, MN/m^2; P = load at rupture, MN; L = support span, m; b = specimen width, m; and d = specimen thickness, m.

ANSI C-64.1—1970, par. 4.6 — Specifies a square parallelapiped either 5.08 mm (0.200 in) or 3.18 mm (0.125 in.) on a side. Single-point central loading is used with the support span being 3.05 mm (1.20 in.) or 1.524 mm (0.60 in.) for the larger and smaller specimen, respectively. Strength is calculated

$$S = 3PL/2bd^2$$

where the terms have the same meaning as above.

The ANSI method gives "strength" values 10 to 25% higher than the ASTM method.

IX. HARDNESS, ROCKWELL

ASTM C-748 — This method measures the resistance to indentation by a 6.35-mm (0.250-in.) ball under a 60-kg major load. Values are reported on the Rockwell L scale and should lie between 0 and 120, with increasing values representing increased resistance to indentation. Technically, the method is limited to fine-grain graphites having a maximum particle size less than 0.80 mm.

X. HARDNESS, SCLEROSCOPE C

ASTM C-886 — This test method requires a Shore Model C-2 Scleroscope. A calibrated, diamond-tipped "hammer" weighing approximately 2.2 g is dropped 25.1 cm (9.89 in.) down a suitably marked vertical glass tube, striking the carbon part and rebounding. Rebound height is visually observed. The hammer must be specifically calibrated for use with carbon and can be used only in the instrument for which it was calibrated. The carbon specimen tested must be rigidly supported, and the surface struck must be flat and normal to the tube axis. The specimen must be dry and the test surface smooth and dustfree to avoid false readings. Specimens thinner than 3 mm (1/8 in.) tend to give high readings. Repeated readings by a skilled operator show a standard deviation of less than two units.

ANSI C 64.1—1970, par. 4.3 — Similar to ASTM C-886. There is no correlation between Scleroscope hardness and Rockwell hardness.

XI. MODULUS OF ELASTICITY

ASTM C-747 — This method is commonly used to determine the dynamic modulus of elasticity at room temperature. The modulus is calculated from the measured fundamental frequency of a long slender rod or prism.

ASTM C-769 — This method measures the speed of sound as it passes down a parallelapiped at least 13 mm on a side and 26 mm long. It is limited to carbons having a grain size smaller than 0.80 mm, and it requires values for the density and Poisson's ratio in order to calculate modulus.

XII. MOISTURE

ASTM C-562 — Uses a ground sample (through 60 mesh) weighing 50 to 60 g and dries it at 105 ± 2°C(221 ± 4°F) in an air convection oven.

XIII. OPEN POROSITY

The three test methods vary principally in the fluid used to fill the pores.

ASTM C-373 — Dry weight is taken after a bake out at 150°C (302°F), then the specimen is impregnated with water and weighed again. Impregnation is achieved by boiling for 5 hr followed by a 24-hr soak in an open beaker.

ANSI C-64.1—1970, par. 4.10 — Impregnation is with SAE-10 petroleum oil. The specimen is submerged in oil at 70°C (158°F) for 3 to 5 hr while the pressure above the oil surface is maintained at 2.67 to 8.00 kN/m^2 (20 to 60 mm Hg absolute) to remove entrapped air. Without removing the specimen, the pressure is raised to atmospheric and the temperature dropped at 20°C (68°F).

ANSI C-64.1—1970, par. 4.12 — Uses n-butanol, $CH_3 (CH_2)_3OH$, as the impregnant. The volume of the carbon body exclusive of open pores is determined by displacement of butanol from a pycnometer of known capacity.

The carbon weight divided by this volume is called its "real density". Porosity (open) is calculated:

$$P = \frac{(RD - AD) \times 100}{RD}$$

where P = porosity (open), volume %; RD = real density, g/cc; and AD = apparent density, g/cc.

XIV. TENSILE STRENGTH

ASTM C-565 — Uses a test specimen 127 mm (5 in.) long with heads 25.4 mm (1 in.) in diameter. The central 70% of the round specimen has a continuous radius profile to avoid stress concentration due to sudden changes in section. The minimum diameter of the central gage section is 7.87 mm (0.310 in.).

ASTM C-749 — Uses a round specimen 120 mm (4 3/4 in.) to 365 mm (14 3/8 in.) long depending upon grain size of the carbon. The gage section is cylindrical, which permits the measurement of strains and the calculation of modulus and Poisson's ratio. The test equipment is more complicated than that of ASTM C-565.

Both tensile test methods have a precision of 5% or better.

XV. THERMAL DIFFUSIVITY

ASTM C-714 — A graphite disc 2 to 4 mm thick and 6 to 12 mm in diameter is subjected to a high-intensity thermal pulse on one face. The temperature of the rear surface as a function of time is recorded on an oscilloscope. Data from this curve together with the disc thickness are used to calculate thermal diffusivity.

XVI. THERMAL EXPANSION

See Séction IV, Coefficient of Thermal Expansion.

XVII. OTHER TESTS

This is by no means a complete listing of test methods used to determine the properties of manufactured carbon and graphite. In the U.S., two organizations — American Society for Testing and Materials and National Electrical Manufacturers' Association — are active in developing new methods. The latest publications of these two organizations should be consulted for test methods developed after 1975.

REFERENCES

1. 1975 Annual Book of ASTM Standards, Part XVII, *American Society for Testing and Materials*, Philadelphia, April 1975.
2. Brushes for Electrical Machines, NEMA CB 1-1970, ANSI C-64.1 — 1970, American National Standards Institute, New York, March 13, 1970.

Appendix C

CHEMICAL COMPATABILITY OF MANUFACTURED CARBONS USED IN MECHANICAL SEALS

Chemical	P-29[a]	P-5[a]	P-03[a]	P-03XHT[a]	PE-6923[a]
Acetaldehyde	+	+	+	+	+
Acetanilide	0	+	+	Δ	+
Acetates, esters, petroleum derivatives	+	+	+	+	+
Acetate solvents	Δ	+	+	Δ	+
Acetic anhydride	+	+	+	+	+
Acetone	+	+	+	Δ	+
Acetone cyanohydrin	+	+	+	Δ	+
Acetophenone	+	+	+	Δ	+
Acetylene gas	Δ	+	+	+	+
Acid					
Abietic	+	+	+	Δ	+
Acetic	+	+	+	0	+
Acetic propionic	+	+	+	0	+
Acetylsalicylic	+	+	+	0	+
Alkyl-arylsulphonic	+	+	+	0	+
Aqua regia	0	0	0	0	Δ
Arsenic	Δ	+	+	0	+
Arsenic, ortho-	Δ	+	+	0	+
Ascorbic	+	+	+	0	+
Battery	Δ	Δ	Δ	0	+
Benzensulfonic	+	+	+	0	+
Benzoic	+	+	+	0	+
Boiler phosphates	+	+	+	0	+
Boric (aqueous)	+	+	+	0	+
Boric (boracic)	+	+	+	0	+
Butyric (butanoic)	+	+	+	Δ	+
Carbolic (phenol	Δ	+	+	Δ	+
Carbonic	+	+	+	0	+
Chloric	0	Δ	+	0	+
Chlorous	0	Δ	+	0	+
Chloroacetic	+	+	+	0	+
Chlorosulfonic	Δ	+	+	0	+
Chromic	0	0	Δ	0	+
Citric	+	+	+	0	+
Coconut, fatty	+	+	+	+	+
Cresylic	Δ	+	+	0	+
Crotonic	+	+	+	Δ	+
Cyanic	+	+	+	0	+
Dichloroacetic	+	+	+	Δ	+
Ethonic	+	+	+	0	+
Fatty	+	+	+	Δ	+
Fluoboric	Δ	+	+	0	+
Fluosilicic	Δ	Δ	+	0	+
Formic	+	+	+	0	+
Fruit	+	+	+	0	+
Fumaric	+	+	+	0	+
Gallic	+	+	+	0	+

Chemical	P-29*	P-5*	P-03*	P-03XHT*	PE-6923*
Fumaric	+	+	+	0	+
Gallic	+	+	+	0	+
Glutamic	+	+	+	0	+
Glutaric	+	+	+	0	+
Hydrobromic	+	Δ	Δ	0	+
Hydrochloric	+	+	+	0	+
Hydrocyanic	+	+	+	0	+
Hydrofluoric	+	Δ	+	0	+
Hydrofluosilicic	Δ	Δ	+	0	+
Hydrogen peroxide	Δ	0	Δ	Δ	+
Hydroxyacetic	+	+	+	0	+
Hypochlorous	Δ	Δ	+	0	+
Isophthalic	+	+	+	Δ	+
Lactic	+	+	+	0	+
Lauric	+	+	+	Δ	+
Levulinic	+	+	+	0	+
Malic	+	+	+	0	+
Malonic	+	+	+	0	+
Mine water	+	+	+	0	+
Mixed sulfuric and nitric	0	0	Δ	Δ	+
Monochloro-acetic	Δ	+	+	0	+
Muriatic	+	+	+	0	+
Naphthenic	+	+	+	Δ	+
Nitric	Δ	0	Δ	0	+
Nitric, fuming	0	0	Δ	0	+
Nitrous	+	Δ	+	0	+
Nordhausen	Δ	Δ	Δ	0	+
Oleic	+	+	+	Δ	+
Oleum	0	0	Δ	Δ	+
Ortho-phosphoric	+	+	+	Δ	+
Organic	+	+	+	Δ	+
Oxalic	+	+	+	+	+
Palmitic	+	+	+	+	+
Perchloric	0	0	0	0	Δ
Phenolacetic	+	+	+	+	+
Phenolsulfonic	+	+	+	+	+
Phosphoric	+	+	+	0	+
Phthalic	+	+	+	Δ	+
Picric	Δ	0	Δ	Δ	+
Propionic	+	+	+	Δ	+
Prussic	+	+	+	Δ	+
Pyrogallic	+	+	+	+	+
Pyroligneous	+	+	+	0	+
Salicic	+	+	+	0	+
Salicylic	+	+	+	0	+
Sorbic	+	+	+	0	+
Stearic (octodecanoic)	+	+	+	Δ	+
Stearic and oleic	+	+	+	Δ	+
Succinic	+	+	+	Δ	+
Sulfuric (0—75%)	Δ	Δ	+	0	+
Sulfuric (75—96%)	0	0	Δ	Δ	+
Sulfuric (fuming)	0	0	Δ	Δ	+
Sulfurous	+	+	+	0	+

Chemical	P-29*	P-5*	P-03*	P-03XHT*	PE-6923*
Tannic	+	+	+	0	+
Tartaric	+	+	+	0	+
Terephthalic	+	+	+	Δ	+
Toluenesulfonic	+	+	+	0	+
Toluic	+	+	+	Δ	+
Trichloroacetic	+	+	+	Δ	+
Uric	+	+	+	0	+
Uric (methyl)	+	+	+	0	+
Valeric	+	+	+	Δ	+
Vinyl acetate	+	+	+	0	+
Acrolein	+	+	+	Δ	+
Acrylonitrile	+	+	+	+	+
Air, dry	+	+	+	+	+
Air, humid	+	+	+	Δ	+
Alcohol					
Allyl	+	+	+	Δ	+
Butyl	+	+	+	Δ	+
Ethyl	+	+	+	Δ	+
Grain	+	+	+	Δ	+
Iso-butyl	+	+	+	Δ	+
Iso-propyl	+	+	+	Δ	+
Methyl	+	+	+	Δ	+
Octyl	+	+	+	Δ	+
Alcohol and lubricating oil	+	+	+	Δ	+
Aldehyde	+	+	+	Δ	+
Ale	+	+	+	0	+
Alkali cleaner	0	+	+	0	+
Alkanes	+	+	+	+	+
Alkyd resins	+	+	+	+	+
Alkyl-arylsulfonics	+	+	+	Δ	+
Alkylaryl sulfonates	+	+	+	Δ	+
Alkylate (light)	+	+	+	+	+
Allyl acetone	+	+	+	Δ	+
Allyl amine	0	+	+	Δ	+
Allyl benzene	+	+	+	+	+
Allyl chloride	+	+	+	+	+
Almond oil	+	+	+	Δ	+
Alpha picoline	+	+	+	Δ	+
Alum (ammonia)	Δ	+	+	0	+
Alum (crhome)	Δ	+	+	0	+
Alum (potash)	+	+	+	0	+
Aluminum acetate	+	+	+	0	+
Aluminum chloride	Δ	Δ	Δ	0	+
Aluminum chlorate	0	0	Δ	0	+
Aluminum hydroxide	Δ	+	+	0	+
Aluminum nitrate	Δ	+	+	0	+
Aluminum sulfate	+	+	+	0	+
Amine	0	+	+	Δ	+
Amine, diethanol	0	+	+	0	+
Amine, fat condensate	Δ	+	+	Δ	+
Ammonia aqueous	0	+	+	0	+

Chemical	P-29[a]	P-5[a]	P-03[a]	P-03XHT[a]	PE-6923[a]
Ammonia, gas (dry)	0	Δ	Δ	Δ	+
Ammonia, gas (wet)	0	+	+	0	+
Ammonia, liquid	0	+	+	Δ	+
Ammonia and oil	Δ	+	+	Δ	+
Ammoniacal liquor	0	+	+	0	+
Ammonium acetate	Δ	+	+	0	+
Ammonium bicarbonate	+	+	+	0	+
Ammonium bifluoride	+	+	+	0	+
Ammonium bromide	+	+	+	0	+
Ammonium carbonate	Δ	+	+	0	+
Ammonium chloride	+	+	+	0	+
Ammonium hydroxide (aqueous)	0	+	+	0	+
Ammonium nitral (75%)	Δ	+	+	0	+
Ammonium nitrate	Δ	Δ	Δ	0	+
Ammonium phosphate, mono- (aqueous)	+	+	+	0	+
Ammonium phosphate, di- (aqueous)	+	+	+	0	+
Ammonium phosphate, tri- (aqueous)	+	+	+	0	+
Ammonium sulfate, (aqueous)	+	+	+	0	+
Ammonium sulphide	Δ	+	+	0	+
Ammonium thiocynate	+	+	+	0	+
Ammonium thiocyanide	Δ	+	+	0	+
Amyl acetate	+	+	+	+	+
Amyl alcohol	+	+	+	Δ	+
Amyl amines	Δ	+	+	0	+
Amyl benzoate	+	+	+	Δ	+
Amyl butyrate	+	+	+	Δ	+
Amyl formate	+	+	+	Δ	+
Amyl nitrate	+	+	+	Δ	+
Amyl phthalate	+	+	+	Δ	+
Amyl propionate	+	+	+	Δ	+
Angamol oil	+	+	+	Δ	+
Aniline	Δ	+	+	0	+
Aniline hydrochloride	+	+	+	0	+
Anthracene	+	+	+	+	+
Antibiotic fermentation broth (without solvent)	Δ	+	+	0	+
Antibiotic fermentation broth (with solvent)	Δ	+	+	0	+
Antifreeze	+	+	+	0	+
Argon gas	+	Δ	Δ	+	+
Arochlor 1248	+	+	+	+	+
Aromatic fuels, <20% aromatic	+	+	+	+	+
Aromatic fuels, >20% aromatic	+	+	+	+	+
Arsenic trichloride	+	+	+	+	+
Asphalt	+	+	+	+	+
Aspirin	+	+	+	Δ	+
ASTM oil #1	+	+	+	+	+
ASTM oil #2	+	+	+	+	+
ASTM oil #3	+	+	+	+	+
ASTM oil #4	+	+	+	+	+
Aviation gasoline	+	+	+	+	+
Baking soda	+	+	+	Δ	+

Chemical	P-29*	P-5*	P-03*	P-03XHT*	PE-6923*
Barium carbonate	+	+	+	+	+
Barium chloride (aqueous)	+	+	+	0	+
Barium hydroxide	Δ	+	+	0	+
Barium nitrate (Aqueous)	+	+	+	0	+
Barium sulfate	+	+	+	+	+
Barium sulfide	0	+	+	0	+
Bauxite	+	+	+	+	+
Beer	+	+	+	0	+
Beer wort	+	+	+	0	+
Beet juice	+	+	+	0	+
Beet pulp	+	+	+	0	+
Beet sugar	+	+	+	Δ	+
Beet sugar liquors	+	+	+	0	+
Benzaldehyde	+	+	+	Δ	+
Benzene	+	+	+	+	+
Benzene (aliphatic)	+	+	+	+	+
Benzine	+	+	+	Δ	+
Benzol	+	+	+	+	+
Bichloride of mercury	+	+	+	0	+
Bittern	+	+	+	0	+
Black liquor	+	+	+	0	+
Bleaching powder	0	Δ	+	0	+
Bleach solution	0	Δ	+	0	+
Blood	+	+	+	0	+
Boiler feed water solutions	Δ	+	+	0	+
Bonderite solutions	+	+	+	0	+
Borax solution	Δ	+	+	0	+
Boron trichloride	+	+	+	Δ	+
Boron trifluoride	+	+	+	Δ	+
Bright copper plating solution	Δ	+	+	0	+
Bright nickle plating solution	Δ	+	+	0	+
Brine calcium chloride	+	+	+	0	+
Brine calcium and magnesium chloride	+	+	+	0	+
Brine calcium and sodium chloride	+	+	+	0	+
Brine chloride	+	.+	+	0	+
Brine, sea water	+	+	+	0	+
Brine, sodium chloride	+	+	+	0	+
Bromine gas	Δ	0	Δ	Δ	+
Bromine (wet	Δ	Δ	Δ	0	+
Bromo methane	+	+	+	Δ	+
Bunker "C" fuel oil	+	+	+	+	+
Butadiene	+	+	+	+	+
Butane (LPG)	+	+	+	+	+
Butanol	+	+	+	Δ	+
Butter	+	+	+	0	+
Buttermilk	+	+	+	0	+
Butyl acetate	+	+	+	Δ	+
Butyl amines	0	+	+	0	+
Butyl benzoate	+	+	+	Δ	+
Butyl butyrate	+	+	+	+	+
Butyl "cellosolve"	+	+	+	Δ	+
Butyl formate	+	+	+	Δ	+

Chemical	P-29*	P-5*	P-03*	P-03XHT*	PE-6923*
Butyl phosphate	+	+	+	+	+
Butyl phthalate	+	+	+	+	+
Butyl propionate	+	+	+	Δ	+
Butylene (butene)	+	+	+	+	+
Calcium bisulfide	Δ	+	+	0	+
Calcium bisulfite	Δ	+	+	0	+
Calcium carbonate (argonite)	+	+	+	0	+
Calcium chlorate	+	Δ	+	0	+
Calcium chloride	+	+	+	0	+
Calcium hydroxide	Δ	+	+	0	+
Calcium hypochlorite	0	Δ	+	0	+
Calcium nitrate	+	+	+	0	+
Calcium oxide	Δ	+	+	0	+
Calcium phosphate	+	+	+	0	+
Calcium sulphate	+	+	+	0	+
Calgon (sodium hexametaphosphate)	Δ	+	+	0	+
Caliche liquors	+	+	+	0	+
Camphor	+	+	+	+	+
Candelilla wax	+	+	+	+	+
Cane sugar juice	+	+	+	0	+
Cane sugar liquors	+	+	+	0	+
Carbinal	+	+	+	0	+
Carbitols	+	+	+	0	+
Carbonate of soda	0	+	+	0	+
Carbon bisulfide	+	+	+	Δ	+
Carbon dioxide (dry)	+	+	+	+	+
Carbon dioxide (wet)	+	+	+	0	+
Carbon disulfide	+	+	+	Δ	+
Carbon Monoxide	+	+	+	+	+
Carbon tetrachloride (dry)	Δ	+	+	+	+
Carbon tetrachloride (wet)	Δ	+	+	0	+
Casein	+	+	+	0	+
Castor oil	+	+	+	Δ	+
Catsup	+	+	+	0	+
Caustic (chloride of sodium)	0	+	+	0	+
Caustic cyanogen	+	+	+	Δ	+
Caustic manganese	+	+	+	0	+
Caustic potash (aqueous)	0	+	+	0	+
Caustic soda (aqueous)	0	+	+	0	+
Caustic strontia (aqueous)	Δ	+	+	0	+
Caustic sulfide	Δ	+	+	0	+
Caustic zinc chloride (aqueous)	Δ	Δ	+	0	+
Cellosolve	+	+	+	Δ	+
Cheeses	+	+	+	0	+
Chinawood oil	+	+	+	Δ	+
Chloroacetaldehyde	+	+	+	0	+
Chloral	+	+	+	Δ	+
Chlorox	0	Δ	+	0	+
Chloride of lime	+	+	+	0	+
Chloride of zinc	Δ	+	+	0	+
Chlorinated solvents (dry)	Δ	+	+	+	+
Chlorinated solvents (wet)	Δ	+	+	0	+

Chemical	P-29®	P-5®	P-03®	P-03XHT®	PE-6923®
Chlorine (dry)	Δ	+	+	+	+
Chlorine (wet)	Δ	Δ	+	0	+
Chloroacetone	+	+	+	Δ	+
Chlorobenzene	+	+	+	+	+
Chlorobenzene, di-	Δ	+	+	+	+
Chlorobenzene, tri-	Δ	+	+	+	+
Chloroethylene	Δ	+	+	+	+
Chloroform	Δ	+	+	+	+
Chloromethyl ether	+	+	+	+	+
Chloropicrin	+	+	+	+	+
Chlorosilane	+	+	+	+	+
Chlorothene	Δ	+	+	+	+
Chocolate	+	+	+	Δ	+
Chrome alum	+	+	+	0	+
Chromic oxide (aqueous)	Δ	Δ	+	0	+
Chromium potassium sulfate (aqueous)	+	+	+	0	+
Cider	+	+	+	0	+
Citrus juices	+	+	+	0	+
Clay slurry	+	+	+	0	+
Coal gas	+	+	+	+	+
Coal tar	+	+	+	+	+
Cocoa butter	+	+	+	+	+
Coconut oil	+	+	+	+	+
Cod liver oil	+	+	+	+	+
Coffee	+	+	+	0	+
Cola drinks	+	+	+	0	+
Condensate water	+	+	+	0	+
Cooling tower water	+	+	+	0	+
Copper acetate (blue verdigris)	+	+	+	0	+
Copper ammonium acetate (aqueous)	+	+	+	0	+
Copper chloride	+	+	+	0	+
Copper cyanide	Δ	+	+	0	+
Copper nitrate	+	+	+	0	+
Copper strike plating solution	Δ	+	+	0	+
Copper sulfate (blue vitriol)	+	+	+	0	+
Copperas (green)	+	+	+	0	+
Core oils	+	+	+	Δ	+
Corn oil	+	+	+	Δ	+
Corn starch	+	+	+	0	+
Corn syrup	+	+	+	0	+
Cotton seed oil	+	+	+	Δ	+
Creosote	+	+	+	+	+
Creosols	+	+	+	+	+
Crotonaldehyde	+	+	+	+	+
Crude oil	+	+	+	+	+
Cryogenics	+	0	0	+	+
Cumene	+	+	+	+	+
Cupric sulfate	+	+	+	0	+
Cutting oil	+	+	+	Δ	+
Cyanide	0	+	+	0	+
Cyanogen	+	+	+	+	+

Chemical	P-29*	P-5*	P-03*	P-03XHT*	PE-6923*
Cyclohexane (hexahydrobenzene)	+	+	+	+	+
Cyclohexanol	+	+	+	Δ	+
Cyclohexene (tetrahydrobenzene)	+	+	+	+	+
Cyclohexylamine	+	+	+	Δ	+
Cycloxexanone	+	+	+	Δ	+
DDT solution (toluene solvent)	+	+	+	+	+
DDT solution (kerosene solvent)	+	+	+	+	+
De-butanizer reflux	+	+	+	+	+
De-ethanizer charge	+	+	+	+	+
De-propanizer reflux	+	+	+	+	+
Detergents	+	+	+	0	+
Dextrin	+	+	+	Δ	+
Dextrose	+	+	+	Δ	+
Diacetone	+	+	+	Δ	+
Diacetone alcohol	+	+	+	Δ	+
Dibutyl amine	Δ	+	+	Δ	+
Dibutyl cellosolve adipate	+	+	+	+	+
Dibutyl phosphite	+	+	+	+	+
Dibutyl phthalate	+	+	+	+	+
Dibromoethyl benzene	+	+	+	+	+
Dibutyl ether	+	+	+	+	+
Dichlorobenzene	Δ	+	+	+	+
Dichloroethane	Δ	+	+	+	+
Dichloropentane	Δ	+	+	+	+
Dichlorohydrin	+	+	+	Δ	+
Diesel fuel	+	+	+	+	+
Diethanol amine	0	+	+	0	+
Diethylamine	0	+	+	0	+
Diethylaniline	Δ	+	+	0	+
Diethylbenzene	+	+	+	+	+
Diethylcarbonate	+	+	+	Δ	+
Diethyl ether	+	+	+	+	+
Diethylene glycol	+	+	+	Δ	+
Diethyl sulfate	+	+	+	Δ	+
Diethylene triamine	0	+	+	0	+
Diethyl maleate	+	+	+	+	+
Diethyl phthalate	+	+	+	+	+
Di-isobutyl ketone	+	+	+	Δ	+
Di-isopropyl ketone	+	+	+	Δ	+
Dimethyl formaldehyde	+	+	+	Δ	+
Dimethyl hydrazine	Δ	+	+	Δ	+
Dimethyl phthalate	+	+	+	+	+
Dimethyl terephthalate	+	+	+	+	+
Dinitrochloro benzene	+	Δ	+	+	+
Dinitrochloro benzene and styrene	+	+	+	+	+
Dinitrotoluene	+	Δ	+	+	+
Dioctyl-amine	0	+	+	Δ	+
Dioctyl-phthalate	+	+	+	+	+
Dioxane	+	+	+	Δ	+
Dipentene	+	+	+	+	+
Diphenyl	+	+	+	+	+
Dish water	+	+	+	0	+

Chemical	P-29*	P-5*	P-03*	P-03XHT*	PE-6923*
Disodium phosphate	Δ	+	+	0	+
Doctor's solution	Δ	+	+	0	+
Dow Corning silicone fluid	+	+	+	+	+
Dowtherm	Δ	+	+	+	+
Dowtherm A	Δ	+	+	+	+
Dowtherm E	Δ	+	+	+	+
Drilling mud	+	+	+	0	+
Dye wood liquors	+	+	+	0	+
Eggs	+	+	+	0	+
Electrolytic cell liquors	+	+	+	0	+
Enamel	+	+	+	Δ	+
Epichlorohydrin	+	+	+	Δ	+
Ester-type plasticizers	+	+	+	+	+
Ethane (methylmethane)	+	+	Δ	+	+
Ethanol (ethyl alcohol)	+	+	+	Δ	+
Ethanol amine, mono-	0	+	+	0	+
Ethanol amine, di-	0	+	+	0	+
Ethanol amine, tri-	0	+	+	0	+
Ether	+	+	+	Δ	+
Ether — diethyl	+	+	+	Δ	+
Ether — petroleum	+	+	+	+	+
Ethyl acetate	+	+	+	Δ	+
Ethyl acrylate	+	+	+	+	+
Ethyl alcohol	+	+	+	Δ	+
Ethyl benzene	+	+	+	+	+
Ethyl benzoate	+	+	+	0	+
Ethyl bromide	+	+	+	+	+
Ethyl butyrate	+	+	+	+	+
Ethyl cellosolve	+	+	+	Δ	+
Ethyl cellulose	+	+	+	Δ	+
Ethyl chloride	Δ	+	+	+	+
Ethyl chloroacetate	+	+	+	Δ	+
Ethyl chlorocarbonate	+	+	+	Δ	+
Ethyl chloroformate	+	+	+	Δ	+
Ethyl chlorohydrin	+	+	+	0	+
Ethyl dichloride	Δ	+	+	+	+
Ethyl ether	+	+	+	Δ	+
Ethyl formate	+	+	+	Δ	+
Ethyl hexanol	+	+	+	Δ	+
Ethyl mercaptan	+	+	+	+	+
Ethyl phthalate	+	+	+	+	+
Ethyl propionate	+	+	+	+	+
Ethyl pyridine	+	+	+	Δ	+
Ethyl sulfate	+	+	+	Δ	+
Ethylene (ethene)	+	+	Δ	+	+
Ethylene bromide	+	+	+	+	+
Ethylene chloride	Δ	+	+	+	+
Ethylene chlorohydrin	+	+	+	Δ	+
Ethylene diamine	Δ	+	+	0	+
Ethylene dichloride	Δ	+	+	+	+
Ethylene glycol	+	+	+	0	+

Chemical	P-29*	P-5*	P-03*	P-03XHT*	PE-6923*
Ehtylene oxide	+	+	+	Δ	+
Ethylene oxide and freon 12	+	+	+	Δ	+
Ethylene thiourea	+	+	+	Δ	+
Ethylene trichloride	Δ	+	+	+	+
Ethylethylene	+	+	+	+	+
Fatty acids	+	+	+	Δ	+
Ferric chloride	Δ	Δ	+	0	Δ
Ferric sulfate	+	+	+	0	+
Ferrous chloride	Δ	Δ	Δ	0	+
Ferrous sulfate	Δ	Δ	Δ	0	+
Ferrous sulfate and lime	+	Δ	+	0	+
Film dope	+	+	+	Δ	+
Fish oil	+	+	+	Δ	+
Fluorine	0	Δ	Δ	Δ	+
Fluorolube	+	+	+	+	+
Formaldehyde (menthanol)	+	+	+	Δ	+
Formalin	+	+	+	0	+
Freon 11	+	+	+	+	+
Freon 12	+	+	+	+	+
Freon 13	+	+	+	+	+
Freon 14	+	+	+	+	+
Freon 21	+	+	+	+	+
Freon 22	+	+	+	+	+
Freon 31	+	+	+	+	+
Freon 32	+	+	+	+	+
Freon 112	+	+	+	+	+
Freon 113	+	+	+	+	+
Freon 114	+	+	+	+	+
Freon 115	+	+	+	+	+
Freon 121	+	+	+	+	+
Freons, liquid	+	+	+	+	+
Freons and refrigerator oil	+	+	+	+	+
Fruit juices	+	+	+	0	+
Fuel oil	+	+	+	+	+
Furfural	+	+	+	Δ	+
Furfuryl alcohol	+	+	+	Δ	+
Gas oil	+	+	+	+	+
Gasoline <20% aromatics	+	+	+	+	+
Gasoline >20% aromatics	+	+	+	+	+
Gasoline, automotive	+	+	+	+	+
Gasoline, ethyl	+	+	+	+	+
Gasoline, high-test	+	+	+	+	+
Gasoline, 100 octane	+	+	+	+	+
Gasoline, 130 octane	+	+	+	+	+
Gasoline and hydrogen sulfide	+	+	+	+	+
Gasoline and mercaptan	+	+	+	+	+
Gelatin	+	+	+	0	+
Glaubers salt (sodium sulfate)	+	+	+	0	+
Glucose	+	+	+	0	+
Glue	+	+	+	Δ	+
Glue sizing	+	+	+	Δ	+
Glycerine	+	+	+	Δ	+

Chemical	P-29•	P-5•	P-03•	P-03XHT•	PE-6923•
Glycerol	+	+	+	Δ	+
Glycols	+	+	+	Δ	+
Grape juice	+	+	+	0	+
Grease	+	+	+	+	+
Green sulfate liquors	+	+	+	0	+
Helium	+	+	Δ	+	+
Heptane	+	+	+	+	+
Hexachloroacetone	+	+	+	Δ	+
Hexane	+	+	+	+	+
Hexene (butylethelene)	+	+	+	+	+
Hexone (iso-butyl methyl ketone)	+	+	+	Δ	+
Hexyl alcohol (hexanol)	+	+	+	Δ	+
High-speed copper plating solution	Δ	+	+	0	+
Hops	+	+	+	0	+
Hydrazine	Δ	Δ	+	Δ	+
Hydrazine hydrate	Δ	+	+	Δ	+
Hydrocarbons, light	+	+	+	+	+
Hydrogen	+	+	Δ	+	+
Hydrogen bromide	+	+	+	Δ	+
Hydrogen chloride	+	+	+	Δ	+
Hydrogen cyanide	+	+	+	Δ	+
Hydrogen fluoride	+	+	+	Δ	+
Hydogen peroxide	Δ	Δ	+	Δ	+
Hydrogen sulfide	+	+	+	Δ	+
Hydrogenated fats	+	+	+	+	+
Hydroquinone	Δ	+	+	+	+
Hydroxylamine	0	+	+	Δ	+
Ice cream	+	+	+	0	+
Ink	+	+	+	0	+
Insecticides (aromatic)	+	+	+	+	+
Insecticides (nonaromatic)	+	+	+	Δ	+
Iodine	0	Δ	Δ	Δ	+
Iodoform	Δ	Δ	+	+	+
Iron sulfate	+	Δ	+	0	+
Iso-butane	+	+	+	+	+
Iso-butyl acetate	+	+	+	Δ	+
Iso-butyl alcohol	+	+	+	Δ	+
Iso-butyl methyl ketone	+	+	+	Δ	+
Iso-butylene	+	+	+	+	+
Iso-butyraldehyde	+	+	+	Δ	+
Isododecane	+	+	+	+	+
Iso-octane	+	+	+	+	+
Iso-pentane	+	+	+	+	+
Iso-propanol	+	+	+	Δ	+
Iso-propyl acetate	+	+	+	Δ	+
Iso-propyl acetone	+	+	+	Δ	+
Iso-propyl alcohol	+	+	+	Δ	+
Iso-propyl amine	+	+	+	Δ	+
Iso-propyl chloride	+	+	+	+	+
Iso-propyl ether	+	+	+	Δ	+
Jelly	+	+	+	0	+
Jet fuel, JP-3	+	+	+	+	+

Chemical	P-29•	P-5•	P-03•	P-03XHT•	PE-6923•
Jet fuel, JP-4	+	+	+	+	+
Jet fuel, JP-5	+	+	+	+	+
Jet fuel, JP-6	+	+	+	+	+
Jet fuel, JP-X	+	+	+	+	+
Kaolin slip suspension in water	+	+	+	0	+
Kerosene	+	+	+	+	+
Ketchup	+	+	+	0	+
Ketones	+	+	+	Δ	+
Krypton	+	Δ	Δ	+	+
Lacquer	+	+	+	+	+
Lacquer solvents	+	+	+	+	+
Lard	+	+	+	Δ	+
Latex	+	+	+	0	+
Lead (molten)	0	+	+	+	0
Lead acetate	+	+	+	0	+
Lead carbonate	+	+	+	0	+
Lead nitrate	+	+	+	0	+
Lead sulfamate	+	+	+	Δ	+
Light petroleum gas	+	+	+	+	+
Ligroin (petroleum ether)	+	+	+	+	+
Lime bleach	Δ	+	+	0	+
Lime slurries	+	+	+	0	+
Lime water	+	+	+	0	+
Lindol (tricresyl phosphate)	+	+	+	+	+
Linseed oil	+	+	+	Δ	+
Liquid oxygen	Δ	0	Δ	+	+
Liquid petroleum gas (LPG)	+	+	+	+	+
Liquor, black	+	+	+	0	+
Liquor, green	Δ	+	+	0	+
Liquor, lime	Δ	+	+	0	+
Liquor, pulpmill	Δ	+	+	0	+
Liquor, steep	+	+	+	0	+
Liquor, sulfate	Δ	+	+	0	+
Liquor, sulfite	Δ	+	+	0	+
Liquor, white	Δ	+	+	0	+
Lissapol paste	+	+	+	Δ	+
Lithium bromide brine	+	+	+	0	+
Lithium carbonate	Δ	+	+	0	+
Lithium chloride brine	+	+	+	0	+
Lithium hydroxide	0	+	+	0	+
Lubricating oil	+	+	+	+	+
Lye, caustic	0	+	+	0	+
Lye, salty	0	+	+	0	+
Magnesium acetate	+	+	+	0	+
Magnesium chloride (bischofite)	+	+	+	0	+
Magnesium hydroxide (brucite)	+	+	+	0	+
Magnesium nitrate	Δ	Δ	+	0	+
Magnesium oxide	+	+	+	Δ	+
Magnesium sulfate (Epsom salt)	+	+	+	0	+
Magnesium sulfite	+	+	+	0	+
Maleic anhydride	+	+	+	Δ	+
Maleic hydrazide	Δ	+	+	+	+

Chemical	P-29*	P-5*	P-03*	P-03XHT*	PE-6923*
Malt	+	+	+	0	+
Manganese chloride	+	+	+	0	+
Manganese sulfate	+	+	+	0	+
Manganous sulfate	+	+	+	0	+
Maple syrup	+	+	+	0	+
Marsh gas (methane)	+	+	+	+	+
Mayonnaise	+	+	+	Δ	+
MEA (mono-ethanol-amine)	0	+	0	0	+
MEA and copper sulfate	0	+	0	0	+
Malamine resins	+	+	+	+	+
Mercaptan	+	+	+	Δ	+
Mercuric chloride	+	+	+	0	+
Mercuric nitrate	+	+	+	0	+
Mercurous nitrate	+	+	+	0	+
Mercury	+	+	+	+	+
Mercury salts	+	+	+	0	+
Mercury vapors	+	+	+	+	+
Mesityl oxide (ketone)	+	+	+	Δ	+
Methane	+	+	Δ	+	+
Methanol	+	+	+	Δ	+
Methyl acetate	+	+	+	+	+
Methyl acrylate	+	+	+	+	+
Methyl alcohol	+	+	+	Δ	+
Methyl benzoate	+	+	+	+	+
Methyl bromide	+	+	+	+	+
Methyl butyl ketone	+	+	+	Δ	+
Methyl butyrate	+	+	+	+	+
Methyl cellosolve	+	+	+	Δ	+
Methyl chloride	Δ	+	+	+	+
Methyl chlorosilanes	Δ	+	+	+	+
Methyl cyclopentane	+	+	+	+	+
Methyl dichloride	Δ	+	+	+	+
Methyl ether	+	+	+	Δ	+
Methyl ethyl ether	+	+	+	Δ	+
Methyl ethyl ketone (MEK)	+	+	+	Δ	+
Methyl formate	+	+	+	Δ	+
Methyl isobutyl ketone	+	+	+	Δ	+
Methyl isopropyl ketone	+	+	+	Δ	+
Methyl isophthalate	+	+	+	+	+
Methyl methacrylate	+	+	+	+	+
Methyl phthalate	+	+	+	+	+
Methyl propionate	+	+	+	Δ	+
Methyl salicylate	+	+	+	+	+
Methylene chloride	Δ	+	+	+	+
Methylene dichloride	0	+	+	+	+
Milk	+	+	+	+	+
Milk of lime slurry	+	+	+	+	+
MIL-F-25558 (RJ-1)	+	+	+	+	+
MIL-L-7808	+	+	+	+	+
MIL-H-5606 (HFA)	+	+	+	+	+
MIL-H-5606 (J43)	+	+	+	+	+

Chemical	P-29*	P-5*	P-03*	P-03XHT*	PE-6923*
MIL-O-8200 (HYDR)	+	+	+	+	+
MIL-O-8515	+	+	+	+	+
Mineral oil	+	+	+	+	+
Mineral spirits	+	+	+	+	+
Mine water	+	+	+	0	+
Miscella 20% soya oil	+	+	+	Δ	+
Molasses	+	+	+	0	+
Monochlorobenzene	Δ	+	+	+	+
Monoethanolamine	Δ	+	+	Δ	+
Monosodium glutamate (MSG)	+	+	+	0	+
Morphine	Δ	+	+	0	+
Mustard	+	+	+	0	+
Naphtha	+	+	+	+	+
Naphtha (sour)	+	+	+	+	+
Naphthalene	+	+	+	+	+
Natural gas	+	+	+	+	+
Neatsfoot oil	+	+	+	Δ	+
Neon	+	+	Δ	+	+
Nickle acetate	+	+	+	0	+
Nickle chloride	+	+	+	0	+
Nickle nitrate	+	+	+	0	+
Nickle plating solution	Δ	+	+	0	+
Nickle salts	+	+	+	0	+
Nickle sulfate	+	+	+	0	+
Nicotine sulfate	+	+	+	Δ	+
Nitrating acids	0	0	Δ	0	Δ
Nitrobenzene	+	+	+	+	+
Nitrobenzine	+	+	+	+	+
Nitrochloroform	Δ	Δ	+	+	+
Nitroethane	Δ	Δ	+	+	+
Nitromethane	Δ	Δ	+	+	+
Nitropropane	Δ	Δ	+	+	+
Nitric oxide	Δ	+	Δ	+	+
Nitric tetraoxide	0	Δ	+	+	+
Nitrogen gas	+	+	Δ	+	+
Nitrous oxide	Δ	+	+	+	+
Nonyl phenol	+	+	+	+	+
Nylon salt	+	+	+	Δ	+
Oakite	Δ	+	+	0	+
Octyl alcohol	+	+	+	Δ	+
Octyl phenol	+	+	+	+	+
Oil					
Animal, bone	+	+	+	Δ	+
Animal, cod	+	+	+	+	+
Animal, lard	+	+	+	Δ	+
Animal, menhadden	+	+	+	Δ	+
Animal, neatsfoot	+	+	+	Δ	+
Animal, sperm	+	+	+	+	+
Animal, whale	+	+	+	+	+
Asphalt base	+	+	+	+	+
Bunker C	+	+	+	+	+

Chemical	P-29*	P-5*	P-03*	P-03XHT*	PE-6923*
Chinawood	+	+	+	Δ	+
Coal tar	+	+	+	+	+
Coconut	+	+	+	Δ	+
Corn	+	+	+	Δ	+
Cottonseed	+	+	+	Δ	+
Creosote	+	+	+	+	+
Crude	+	+	+	+	+
Cutting	+	+	+	Δ	+
Diesel, #2D	+	+	+	+	+
Diesel, #3D	+	+	+	+	+
Diesel, #4D	+	+	+	+	+
Diesel, #5D	+	+	+	+	+
Domestic fuel	+	+	+	+	+
Edible	+	+	+	Δ	+
Essential	+	+	+	0	+
Fuel #1	+	+	+	+	+
Fuel #2	+	+	+	+	+
Fuel #3	+	+	+	+	+
Fuel #5A	+	+	+	+	+
Fuel #5B	+	+	+	+	+
Fuel #6	+	+	+	+	+
Fusal (alcohol)	+	+	+	0	+
Gas	+	+	+	+	+
Hydraulic	+	+	+	+	+
Insulating	+	+	+	+	+
Jet fuel	+	+	+	+	+
Kerosene	+	+	+	+	+
Lean	+	+	+	+	+
Linseed	+	+	+	Δ	+
Lubricating	+	+	+	+	+
Mineral	+	+	+	+	+
Navy II	+	+	+	+	+
Nut	+	+	+	Δ	+
Olive	+	+	+	Δ	+
Palm	+	+	+	Δ	+
Paraffin base	+	+	+	+	+
Peanut	+	+	+	Δ	+
Pine	+	+	+	Δ	+
Quenching	+	+	+	Δ	+
Rapeseed	+	+	+	Δ	+
Rich	+	+	+	+	+
Silicone	+	+	+	+	+
Soya bean	+	+	+	Δ	+
Transformer	+	+	+	+	+
Turbine, lube	+	+	+	+	+
Ucon	+	+	+	Δ	+
Vegetable	+	+	+	Δ	+
Whale	+	+	+	Δ	+
Oil and ammonia	Δ	+	+	0	+
Oil and freons	+	+	+	+	+
Oil and methylene chloride	Δ	+	+	+	+

Chemical	P-29•	P-5•	P-03•	P-03XHT•	PE-6923•
Oil and sulfur dioxide	+	+	+	+	+
Olefin, crude	+	+	+	+	+
Oleomargarine	+	+	+	Δ	+
Oleums	0	0	Δ	Δ	+
Oleyl alcohol	+	+	+	Δ	+
Orthodichlorobenzene	+	+	+	+	+
O S 45 Type IV	+	Δ	Δ	Δ	+
Oxygen, gas	+	+	Δ	+	+
Oxygen, liquid	Δ	0	Δ	+	+
Paint	+	+	+	Δ	+
Paint vehicles	+	+	+	Δ	+
Paper Stock <1.5% acidic	+	+	+	0	+
Paper Stock <3% acidic	+	+	+	0	+
Paper Stock >3% acidic	+	+	+	0	+
Papter Stock <1% alkaline	+	+	+	0	+
Paper Stock <3% alkaline	Δ	+	+	0	+
Paper Stock >3% alkaline	Δ	+	+	0	+
Para-cymene	+	+	+	+	+
Paraffin oils	+	+	+	+	+
Paradichlorobenzene	Δ	+	+	+	+
Paraformaldehyde	+	+	+	Δ	+
Paraldehyde	+	+	+	Δ	+
Pectin liquor	+	+	+	0	+
Penicillin (liquid)	+	+	+	0	+
Pentachlorophenol	Δ	+	+	Δ	+
Pentane	+	+	+	+	+
Pentane liquid	+	+	+	+	+
Pentasol	+	+	+	Δ	+
Perchlorolthylene	0	+	Δ	+	+
Peroxide of hydrogen	Δ	Δ	Δ	0	+
Petrolatum	+	+	+	+	+
Petroleum ether	+	+	+	+	+
Petroleum jelly	+	+	+	+	+
Petroleum spirit <20% aromatics	+	+	+	+	+
Petroleum spirit >20% aromatics	+	+	+	+	+
Phenol	+	+	+	Δ	+
Phenol and formaldehyde	+	+	+	Δ	+
Phenol and water	+	+	+	0	+
Phenolic resins	+	+	+	Δ	+
Phlorol	Δ	Δ	Δ	Δ	+
Phosgene	+	+	+	+	+
Phosphorous oxychloride	+	+	+	Δ	+
Phosphorous trichloride	0	Δ	+	+	+
Photographic developers	+	+	+	0	+
Phthalic anhydride	+	+	+	+	+
Phthalate esters	+	+	+	+	+
Pickle solutions	+	+	+	0	+
Pine oil	+	+	+	Δ	+
Platicizers (liq. dibutyl cellusolve adipate)	+	+	+	+	+
Platicizers, ester type	+	+	+	+	+
Plating solution, chrome	0	Δ	Δ	0	+

Chemical	P-29*	P-5*	P-03*	P-03XHT*	PE-6923*
Plating solution, copper, acid	+	+	+	0	+
Plating solution, copper cyanide	Δ	+	+	0	+
Plating solution, cyanide	Δ	+	+	0	+
Plating solution, nickle cyanide	Δ	+	+	0	+
Polyglycols	+	+	+	Δ	+
Polyethylene	+	+	+	+	+
Polyphenyl ether	+	+	+	+	+
Polystyrene	+	+	+	+	+
Polyurethane	+	+	+	+	+
Polyvinyl acetate	+	+	+	+	+
Potash sulfide	Δ	+	+	0	+
Potassium alum	+	+	+	0	+
Potassium bicarbonate (aqueous)	+	+	+	0	+
Potassium bichromate (aqueous)	Δ	Δ	+	0	+
Potassium bisulfide	Δ	+	+	0	+
Potassium bromide (aqueous)	+	+	+	0	+
Potassium carbonate (aqueous)	0	+	+	0	+
Potassium chlorate	Δ	0	Δ	0	+
Potassium chloride (aqueous)	+	+	+	0	+
Potassium cyanide (aqueous)	0	+	+	0	+
Potassium dichromate	0	Δ	+	0	+
Potassium hydroxide, <10%	0	+	+	0	+
Potassium hydroxide, >10%	0	+	+	0	+
Potassium iodide	Δ	Δ	+	0	+
Potassium nitrate	Δ	Δ	+	0	+
Potassium perfluoro acetate	+	+	+	0	+
Potassium permanganate	0	Δ	+	0	+
Potassium phosphate, mono-	+	+	+	0	+
Potassium phosphate, di-	+	+	+	0	+
Potassium phosphate, tri-	Δ	+	+	0	+
Potassium silicate	Δ	+	+	0	+
Potassium sulfate (arconite)	+	+	+	0	+
Potassium sulfide	Δ	+	+	0	+
Prestone	+	+	+	Δ	+
Primary flash distillate (PFD)	+	+	+	+	+
Producer gas	+	+	+	+	+
Propane (dimethylmethane)	+	+	Δ	+	+
Propanol	+	+	+	0	+
Propionaldehyde	+	+	+	Δ	+
Propyl acetate	+	+	+	Δ	+
Propyl alcohol	+	+	+	0	+
Propyl benzoate	+	+	+	+	+
Propyl butyrate	+	+	+	+	+
Propyl formate	+	+	+	+	+
Propyl phthalate	+	+	+	+	+
Propyl propionate	+	+	+	+	+
Propylene (propene)	+	+	+	+	+
Propylene carbonate	Δ	+	+	Δ	+
Propylene dichloride	Δ	+	+	+	+
Propylene glycol	+	+	+	Δ	+
Propylene oxide	+	+	+	Δ	+

Chemical	P-29*	P-5*	P-03*	P-03XHT*	PE-6923*
Protein slurry	+	+	+	0	+
Pulp stock	+	+	+	0	+
PVA emulsions	+	+	+	0	+
PVC slurry	Δ	+	+	0	+
Pyridine	Δ	+	+	Δ	+
Raffinate	+	+	+	Δ	+
Resorcinol	+	+	+	Δ	+
Rosin	+	+	+	Δ	+
Saccharine	+	+	+	0	+
Salad oil	+	+	+	Δ	+
Sal ammoniac	+	+	+	0	+
Salt cake (sodium sulfate)	+	+	+	0	+
Salt cake, aqueous	+	+	+	0	+
Salt water	+	+	+	0	+
Saltpeter	Δ	Δ	+	0	+
Sea water	+	+	+	0	+
Sewage	+	+	+	0	+
Shellac	+	+	+	+	+
Silicone fluids	+	+	+	+	+
Silicone oils	+	+	+	+	+
Silicon tetrachloride	+	+	+	Δ	+
Silver nitrate	+	+	+	0	+
Skydrol 500	+	+	+	+	+
Skydrol 7000	+	+	+	+	+
Slop, brewery	+	+	+	0	+
Slop, distillers	+	+	+	0	+
Soap and soap solutions	Δ	+	+	0	+
Soap liquors	Δ	+	+	0	+
Soda ash (sodium carbonate)	0	+	+	0	+
Soda, caustic (sodium hydroxide)	0	+	+	0	+
Sodium acetate	Δ	+	+	0	+
Sodium aluminate	Δ	+	+	0	+
Sodium aluminate lime	0	+	+	0	+
Sodium acid sulfate	+	+	+	0	+
Sodium bicarbonate	+	+	+	0	+
Sodium bichromate	0	+	Δ	0	+
Sodium bisulfate	+	+	+	0	+
Sodium bisulfite	0	+	0	0	+
Sodium borate (Borax)	Δ	+	Δ	0	+
Sodium bromide	+	+	+	0	+
Sodium carbonate (washing soda)	Δ	+	Δ	0	+
Sodium chlorate	0	0	Δ	0	+
Sodium chloride	+	+	+	0	+
Sodium chromate	0	Δ	+	0	+
Sodium cyanamide	Δ	+	+	0	+
Sodium cyanide	0	+	+	0	+
Sodium dichromate	0	Δ	+	0	Δ
Sodium dithionate	Δ	+	+	0	+
Sodium fluoride	Δ	+	+	0	+
Sodium hexametaphosphate	Δ	+	+	0	+

Chemical	P-29•	P-5•	P-03•	P-03XHT•	PE-6923•
Sodium hydrosulfite	Δ	+	+	0	+
Sodium hydroxide	0	+	+	0	+
Sodium hypochlorite	0	Δ	+	0	+
Sodium metaphosphate	0	+	+	0	+
Sodium metasilicate	Δ	+	+	0	+
Sodium nitrate	Δ	Δ	+	0	+
Sodium nitrite	Δ	+	+	0	+
Sodium perborate	0	Δ	+	0	+
Sodium peroxide	Δ	Δ	+	0	+
Sodium phosphate, mono-	Δ	+	+	0	+
Sodium phosphate, di-	0	+	+	0	+
Sodium phosphate, tri-	0	+	+	0	+
Sodium phosphate, meta-	0	+	+	0	+
Sodium plumbite	Δ	+	+	0	+
Sodium silicate	0	+	+	0	+
Sodium sulfate	+	+	+	0	+
Sodium sulfide	0	+	+	0	+
Sodium sulfite	0	+	+	0	+
Sodium tetraborate	0	+	+	0	+
Sodium thiocyanate	Δ	+	+	0	+
Sodium thiosulfate (hydro)	Δ	+	+	0	+
Soft drinks	+	+	+	0	+
Solvasol 1	Δ	+	+	Δ	+
Solvasol 2	Δ	+	+	Δ	+
Solvasol 3	Δ	+	+	Δ	+
Solvasol 73	Δ	+	+	Δ	+
Solvasol 74	Δ	+	+	Δ	+
Sorbitol	+	+	+	Δ	+
Soybean oil	+	+	+	Δ	+
Stannic chloride	+	+	+	0	+
Stannous fluoride	Δ	+	+	0	+
Starch slurry	+	+	+	0	+
Steam	Δ	+	+	0	+
Steep liquors	Δ	+	+	0	+
Stoddard solvent	+	+	+	+	+
Strontium nitrate (aqueous)	Δ	Δ	+	0	+
Styrene (vinylbenzene)	+	+	+	+	+
Styrene (and di-nitrochlorobenzene)	Δ	+	+	+	+
Sugar solutions	+	+	+	0	+
Sulfahydrate	+	+	+	0	+
Sulfate liquors, acid	+	+	+	0	+
Sulfate liquors, base	0	+	+	0	+
Sulfate of lime	+	+	+	0	+
Sulfide of hydrogen	+	+	+	Δ	+
Sulfide of sodium	0	+	+	0	+
Sulfinol	Δ	+	+	Δ	+
Sulfite pulp	+	+	+	0	+
Sulfolignin salts	Δ	+	+	0	+
Sulfonated fatty alcohol	+	+	+	Δ	+
Sulfur, molten	Δ	+	+	+	+

Chemical	P-29*	P-5*	P-03*	P-03XHT*	PE-6923*
Sulfur in water	+	+	+	0	+
Sulfur chloride	Δ	+	+	+	+
Sulfur dioxide (dry)	+	+	+	+	+
Sulfur dioxide (wet)	+	+	+	0	+
Sulfur dioxide in water	+	+	+	0	+
Sulfur trioxide (dry)	Δ	Δ	+	+	+
Sulfur trioxide (wet)	0	0	Δ	0	+
Sulfur chlorohydrin	Δ	+	+	Δ	+
Sylvite	+	+	+	0	+
Syrup	+	+	+	0	+
Tall oil	+	+	+	Δ	+
Tallow	+	+	+	Δ	+
Tanning liquors	+	+	+	0	+
Tar, bituminous	+	+	+	+	+
Tar, hot	+	+	+	+	+
Tar, pine	+	+	+	Δ	+
Tar and ammonia in water	+	+	+	0	+
Teepol	Δ	+	+	Δ	+
Tetrabromoethane	Δ	Δ	+	+	+
Tetrachloroethane	0	+	+	+	+
Tetrachloroethylene	Δ	+	+	+	+
Tetra ethyl lead	+	+	+	+	+
Tetrahydrofuran	+	+	+	Δ	+
Tetramine C	+	+	+	Δ	+
Tetraphenyl	+	+	+	+	+
Therminol #1	Δ	+	+	+	+
Therminol #2	Δ	+	+	+	+
Therminol #3	Δ	+	+	+	+
Thinner for cellulose lacquer	+	+	+	Δ	+
Thinner for oil paint	+	+	+	+	+
Titanium tetrachloride	+	+	+	Δ	+
Toluene (toluol)	+	+	+	+	+
Tomato pulp	+	+	+	0	+
Toothpaste	+	+	+	0	+
Toxaphene	+	+	+	Δ	+
Trichlorhydrin	Δ	+	+	+	+
Trichlorobenzene	Δ	+	+	+	+
Trichloroethane (dry)	Δ	+	+	+	+
Trichloroethane (wet)	Δ	+	+	0	+
Trichloroethylene (dry)	0	+	+	+	+
Trichloroethylene (wet)	0	+	+	0	+
Trichloronitromethane	0	+	+	+	+
Tricresyl phosphate	+	+	+	+	+
Triethanolamine	0	+	+	0	+
Triethylamine	0	+	+	Δ	+
Tri-fluoro-vinyl chloride	Δ	+	+	+	+
Trisodium phosphate	0	+	+	0	+
Turpentine	+	+	+	+	+
Urea, anhydrous (carbamide)	Δ	+	+	0	+
Urea, wet (carbamide)	Δ	+	+	0	+
Urea and phenolic resins	Δ	+	+	Δ	+
Urine	+	+	+	0	+

Chemical	P-29•	P-5•	P-03•	P-03XHT•	PE-6923•
Varnish, aromatic	+	+	+	+	+
Varnish, nonaromatic	+	+	+	+	+
Vegetable juices	+	+	+	0	+
Vegetable oils	+	+	+	Δ	+
Vetrocoke solution (wet)	+	+	+	0	+
Vinegar	+	+	+	0	+
Vinyl acetate	+	+	+	Δ	+
Vinyl chloride	+	+	+	+	+
Vinyl pyridine	+	+	+	+	+
Vitriol, white	0	0	Δ	Δ	+
Water	+	+	+	0	+
Hot	Δ	+	+	0	+
Acid, mine	+	+	+	0	+
Boiler feed	+	+	+	0	+
Brackish	+	+	+	0	+
Chilled air washing	+	+	+	0	+
Condenser service	+	+	+	0	+
Condensate	+	+	+	0	+
Cooling tower	+	+	+	0	+
Deionized	+	+	+	0	+
Demineralized	+	+	+	0	+
Detergent	+	+	+	0	+
Distilled	+	+	+	0	+
Drinking	+	+	+	0	+
Grit carrying	+	+	+	0	+
Heavy	+	+	+	0	+
Sandy	+	+	+	0	+
Sea	+	+	+	0	+
Soapy	+	+	+	0	+
Water and soluble oil	+	+	+	0	+
Water glass	Δ	+	+	0	+
Waxes	+	+	+	+	+
Whale oil	+	+	+	Δ	+
Whey	+	+	+	0	+
Whiskey	+	+	+	0	+
White liquor	+	+	+	0	+
White water, paper mill	+	+	+	0	+
White spirit	+	+	+	0	+
Wine	+	+	+	0	+
Wood pulp	+	+	+	0	+
Wood vinegar	+	+	+	0	+
Wort (beer)	+	+	+	0	+
Xenon	+	+	Δ	+	+
Xylene (dimethylbenzene)	+	+	+	+	+
Xylene (xylole, xylol)	+	+	+	+	+
Xylene, meta-	+	+	+	+	+
Xylene, ortho-	+	+	+	+	+
Xylene, para-	+	+	+	+	+
Yeast	+	+	+	0	+
Yogurt	+	+	+	0	+
Zeolite treated water	+	+	+	0	+

Chemical	P-29*	P-5*	P-03*	P-03XHT*	PE-6923*
Zinc acetate	+	+	+	0	+
Zinc ammonium chloride	+	+	+	0	+
Zinc chloride (dry)	Δ	+	+	+	+
Zinc chloride (wet)	Δ	+	+	0	+
Zinc cyanide	Δ	+	+	0	+
Zinc hydrosulfite	+	+	+	0	+
Zinc nitrate	+	+	+	8	+
Zinc oxide	+	+	+	Δ	+
Zinc phosphate	+	+	+	0	+
Zinc sulfate	+	+	+	0	+

Note: + = compatible, no known interactions;
Δ = conditional, depends on conditions;
0 = incompatible, not recommended.

* Grade designations are those of Pure Carbon Company, Inc., St. Marys, Pa.

INDEX

A

B

C

H

I

J

K

L

M

N

O

P

Q

R

S

T

U

V

W

Y